Energy Infrastructure Protection and Homeland Security

FRANK R. SPELLMAN AND
REVONNA M. BIEBER

GOVERNMENT INSTITUTES
An imprint of
THE SCARECROW PRESS, INC.
Lanham • Toronto • Plymouth, UK
2010

Published by Government Institutes
An imprint of The Scarecrow Press, Inc.
A wholly owned subsidiary of The Rowman & Littlefield Publishing Group, Inc.
4501 Forbes Boulevard, Suite 200, Lanham, Maryland 20706
http://www.govinstpress.com

Estover Road, Plymouth PL6 7PY, United Kingdom

British Library Cataloguing in Publication Information Available

Library of Congress Cataloging-in-Publication Data

Spellman, Frank R.
Energy infrastructure protection and homeland security / Frank R. Spellman and Revonna M. Bieber.
p. cm.
Includes bibliographical references and index.
ISBN 978-1-60590-678-2 (pbk. : alk. paper) — ISBN 978-1-60590-679-9 (electronic)
1. Energy facilities—Protection—United States. 2. Terrorism—United States—Prevention.
3. Civil defense—United States. I. Bieber, Revonna M., 1976– II. Title.

TJ163.25.U6S64 2010
363.325'9333790973—dc22
2010021450

™ The paper used in this publication meets the minimum requirements of American National Standard for Information Sciences—Permanence of Paper for Printed Library Materials, ANSI/NISO Z39.48-1992.

Printed in the United States of America

For JoAnn Garnett-Chapman,
The ultimate energizer

* * *

For Adeline, Tana, and Gibson,
May your future be bright and secure!

Contents

Preface

The fourth of a new Government Institutes series on critical infrastructure and homeland security, *Energy Infrastructure Protection and Homeland Security* is a reference source that serves U.S. energy production/extraction/manufacturing/refining/sales/distribution-transmission businesses and managers who want quick answers to complicated questions—to help employers and employees handle security threats they must be prepared to meet on a daily basis. In the post–September 11 world, the possibility of energy infrastructure terrorism—the malicious use of weapons to cause devastating damage to the energy industrial sector along with its cascading effects—is very real.

In this text, the term *energy* or *energy sector* includes producing, refining, transporting, generating, transmitting, conserving, building, distributing, maintaining, and controlling energy systems and system components. Moreover, because they are important by-products of oil and natural gas, petrochemical security is also addressed herein.

Prior to 9/11, as part of various security consulting assignments primarily based on the requirements of OSHA's Process Safety Management Standard (PSM) and USEPA's Risk Management Program (RMP), we made several collateral assessments of energy infrastructure site security at facilities primarily involved in chemical production, storage, and/or usage. These site audits were comprehensive in nature. From interviews with site managers and security staff personnel, we found the following concerns:

On the one hand

- Energy infrastructure sites were well aware of their vulnerabilities and are conducting significant efforts to increase planning and preparedness.
- Cooperation through industry groups has resulted in substantial information sharing of effective and best practices across the sector.

- Many site owners and operators have extensive experience with infrastructure protection.
- Recently, many site owners and operators have focused their attention on cyber security.

On the other hand

- Overall, security at energy infrastructure locations ranged from fair to very poor.
- Energy infrastructure site security managers were very pessimistic about their ability to deter sabotage by in-house employees, yet none of them had implemented simple background checks for key employees such as energy production operators.
- None of the corporate security staff had been trained to identify combinations of common chemicals at their sites that could be used as improvised explosives and incendiaries.

In one confidential client's report, we noted that

> among the "soft targets" that we identified as potential terrorist sites/targets were hydroelectric dams, electrical distribution power stations and transmission towers, petroleum refineries, fuel transport and end-user distribution sites, the coal industry and its associated ancillaries, the nuclear power industry, the renewable energy industry [wind power (especially wind farms), solar power, and alternative fuel manufacturers and distributors], silviculture (forest) product production (firewood as a source of fuel), along with compressed gases in tanks, pipelines, and pumping stations; and pesticide manufacturing and supply distributors.

Typically, the energy industry comprises (Energy Industry 2009):

- the petroleum industry, including oil companies, petroleum refiners, fuel transport, and end-user sales at gas stations
- the gas industry, including natural gas extraction, and coal gas manufacture, as well as distribution and sales
- the electrical power industry, including electricity generation, electric power distribution, and sales
- the coal industry
- the nuclear power industry
- the renewable energy industry, comprising alternative energy and sustainable energy companies, including those involved in hydroelectric power, wind power, and solar power generation, and the manufacture, distribution, and sale of alternative fuels.
- traditional energy industry based on the collection and distribution of firewood, the use of which, for cooking and heating, is particularly common in poorer countries

In this book, however, we focus on the three interrelated energy infrastructure segments: electricity, petroleum, and natural gas.

This book (and the thirteen forthcoming volumes of the critical infrastructure series) was written as a result of 9/11 to address these concerns. *Energy Infrastructure Protection and Homeland Security*, in particular, was fashioned in response to the critical needs of energy production managers, energy transmission and distribution managers, energy infrastructure engineers, design engineers, process managers at any level of energy production, engineers, environmental professionals, security professionals (physical and cyber security), industrial hygienists, students—and for anyone with a general interest in the security of energy infrastructure systems. It is important to point that our energy infrastructure (as is the case with the other sixteen critical infrastructures) cannot be made absolutely immune to all possible intrusions/attacks; thus, it takes a concerted, well-thought-out effort to incorporate security upgrades in the retrofitting of existing systems and careful security planning for all new energy infrastructure components. These upgrades or design features need to address issues of monitoring, response, critical infrastructure redundancy, and recovery to minimize risk to the facility/infrastructure.

Energy Infrastructure Protection and Homeland Security presents common-sense methodologies in a straightforward, almost blunt manner. Why so blunt? At this particular time, when dealing with security of workers, family members, citizens, and society in general—actually, with our very way of life—politically correct presentations on security might be the norm, might be expected, might be politically correct, and might be demanded. Frankly, however, our view is that there is nothing normal or subtle about killing thousands of innocent people; mass murders certainly should not be expected; and the right to live in a free and safe environment is a reasonable demand.

This text is accessible to those who have no experience with the energy industry and/or homeland security. If you work through the text systematically, you will gain an understanding of the challenge of domestic preparedness—that is, an immediate need for a heightened state of awareness of the present threat facing the energy industrial sector as a potential terrorist target. Moreover, you will gain knowledge of security principles and measures that can be implemented—adding a critical component not only to your professional knowledge but also giving you the tools needed to combat terrorism in the homeland—our homeland.

Frank R. Spellman
Revonna M. Bieber
Norfolk, Virginia

1

Introduction

Open pit coal processing plant, Gillette, Wyoming

Never underestimate the time, expense, blood, sweat, and effort a terrorist will expend to compromise the security of any industrial facility.

U.S. ENERGY INFRASTRUCTURE: KID IN CANDY STORE SYNDROME*

U.S energy infrastructure—it is everywhere, in every direction, in almost endless combinations, interrelationships, and quantities. In making this simple observation, it matters little whether potential terrorists are of foreign or domestic origin. The fact is they only need to look around or surf the Internet to locate the components of the U.S. energy sector. In doing this, terrorists must be in a state of utter fascination, with many good or tempting goodies (targets) all around them.

Let's take a look at what anyone, including terrorists, can find out about our energy infrastructure.

*Information contained in this section is adapted from *America's Energy Infrastructure: A Comprehensive Delivery System.* www.netl.doe.gov/publications/press/2001 (accessed April 16, 2009).

- In the somewhat desolate Columbia-Snake River basin of Washington and Idaho states and surrounding the Hanford Site (a.k.a. Hanford Works, Hanford Project, Hanford Nuclear Reservation), with its nine decommissioned weapons production nuclear reactors and 53 million gal (204,000 m^3) of high-level radioactive waste, are several hydroelectric dams, strung like beads on the threads known as the Columbia and Snake Rivers. These dams consist of Grand Coulee, Chief Joseph, Wells, Rocky Reach, Rock Island, Wanapum, Priest Rapids, and Ice Harbor. Going westward along the Columbia are McNary, John Day, The Dalles, and Bonneville dams. To the east along the Snake River are Lower Monumental, Little Goose, Lower Granite; finally, to the south along the Snake River bordering Idaho-Oregon, Hells Canyon, Oxbow, and Brownlee dams.
- At the present time, 104 licensed nuclear power plants are located in many states in the U.S. The power plants are listed by group depending on the region of the U.S. in which they are located. Region IV is the largest geographical region with the smallest number of power plants, twenty plants. The region includes the states of North Dakota, South Dakota, Nebraska, Missouri, Arkansas, Louisiana, and all the states west of these. Region III includes the states of Minnesota, Wisconsin, Michigan, Iowa, Illinois, Indiana, and Ohio. This region includes twenty-four nuclear power stations. Region II includes the southeastern states of West Virginia, Kentucky, Virginia, Tennessee, North Carolina, South Carolina, Georgia, Alabama, Mississippi, Florida, and the territories of Puerto Rico and the Virgin Islands. This region includes thirty-four nuclear power stations. Region I includes the New England states of Maine, New Hampshire, Vermont, New York, Massachusetts, Pennsylvania, Connecticut, Jew Jersey, Delaware, Maryland, and Washington, DC. This region includes twenty-six nuclear power plants.
- Hydroelectric dams and nuclear power plants produce electrical power, of course. However, keep in mind that they are only the best-known (the crown jewels) and easily recognized producers of electricity. In the U.S., we also generate electrical power at less-known plants that use coal, petroleum liquids, petroleum coke, and natural and other gases.
- The production of electrical energy involves more than just its initial production. Along with generation and production, transmission and distribution of electrical power are also part of the overall production process. Electricity infrastructure includes a North American power grid made up of long-distance transmission lines that move electricity from region to region, as well as the local distribution lines that carry electricity to homes and businesses.
- Petroleum products are pumped from the ground in domestic oil fields and conveyed through gathering lines to pipelines and then to refineries, where it is trans-

formed into gasoline, diesel fuel, or heating oil. These finished products then travel through pipelines and tanker trucks to distribution outlets for purchase by consumers. Finished product pipeline systems involve a complex infrastructure of their own, including compressor stations or plumbing stations and control systems that open and close valves and which produce flow through pipes, often with the use of computer technology.

- Facilities turn raw natural resources into useful energy products.
- Energy infrastructure also includes rail networks, truck lines, and marine transportation.

Again, an individual terrorist or group of terrorists does not have to look far in the U.S. or on the Internet to find the location of energy infrastructure targets. Terrorists also have little trouble in recognizing that energy infrastructure is vulnerable to physical and cyber disruption that could threaten not only its integrity and safety but also the integrity and safety of large regions of the entire country. Moreover, one need not be a rocket scientist to determine that U.S. power infrastructure is complex and interdependent. Consequently, any disruption of one element can have extensive consequences to others, if not all.

WE CALL IT 9/11; OTHERS CALL IT CANTOR FITZGERALD

Early in the morning of August 19, 2000, a Saturday, a thirty-inch-diameter natural gas transmission pipeline operated by El Paso Natural Gas company accidentally ruptured (due to severe internal erosion of a pipe) adjacent to the Pecos River near Carlsbad, New Mexico. The released gas ignited and burned for fifty-five minutes. Twelve persons who were camping under a concrete-decked steel bridge that supported the pipeline across the river were killed and their three vehicles destroyed. Two nearby steel suspension bridges for gas pipelines across the river were extensively damaged. The major safety issues identified in the post investigation were on the design and construction of the pipeline, the adequacy of El Paso Natural Gas Company's internal corrosion control program, the adequacy of federal safety regulations for natural gas pipelines, and the adequacy of federal oversight of the pipeline operation. Other examples of natural phenomena (events) include wildfires, flash floods, earthquakes, active volcanoes, droughts, and storms. These natural events are not entirely predictable, and they cannot yet be controlled or prevented (Meyer 2005).

Accidental incidents involving new types of emergencies began to surface in the 1940s. They were linked with the behavior of certain chemical and energy products (gas, oil, coal, nuclear, LPG, etc.) collectively called hazardous materials whenever they are misused or involved unintended mishaps and fires.

DID YOU KNOW?

Before the first World Trade Center bombing in 1993 and the 1995 Oklahoma City bombing, emergency incidents were primarily or generally thought to be caused by natural or accidental events.

Where there are people, there are energy sources/products—from the busiest metropolis to the most remote village. In their myriad forms, energy is at the heart of our technology-based society. Energy can make our lives longer, healthier, and more productive, prosperous, and comfortable. Energy is essential to the U.S. economy and standard of living; its many forms allow us to live or experience the trappings of the "good life." The energy industry manufactures products that are fundamental elements of other economic sectors. For example, it produces electricity to light our homes and run our machines, energy-supplied heat can warm our homes or fuel our industrial furnaces, and polymers create plastics from petroleum for innumerable household and industrial products. The most common of these include plastics, leather, paper, rubber, paints, textiles, pesticides, solvents, detergents, fuels, medicines, fertilizers, building materials, electronics, sporting equipment, and automobiles. In addition it is important to point out that an enormous amount of the energy industry's products go to industrial applications as well as domestic or household applications. To eliminate energy products from our society would not only be impractical, it would be undesirable. Instead, since it is impractical and undesirable to eliminate energy products from our society, we must find ways to work and live safely with these products.

It was not that far in our distant past that we dealt mostly with the ramifications of natural disasters. Although never welcomed, these natural occurrences were expected, planned for, guarded against, and recovered from. In regard to disasters, they were our primary concern. Of course, there have always been the man-made or human-caused disasters that we also had to contend with. The sinking of the *Titanic* prompted new safety regulations for passenger ships; a fire at New York City's Triangle Shirtwaist Factory in 1911 caused the deaths of 146 young women and was the catalyst for the development of workplace safety and fire regulations. However, the ultimate wake-up call for Americans came in the form of the events that we now summarize as "9/11" or, as some characterize it, Cantor Fitzgerald 9/11 (because of the 658 employees of this business firm who were killed in the World Trade Center on 9/11/2001). Call it what you will; one fact remains, however: on September 11, 2001, terrorists struck at the heart of America—on American soil and in a way that is unforgettable. Airplanes filled with people and fuel were turned into guided missiles of death and destruction.

What the 9/11 terrorists provided us with was a wake-up call. It was a wake-up call that alerted us to a new type of hateful venom—coldly delivered by groups or individuals to kill massive numbers of people, cause substantial property damage, and affect economic stability. Most importantly, for the public and emergency responders the events of 9/11 brought to the forefront knowledge and awareness that just about anything unthinkable is possible. The bottom line: security of the homeland is vital to all of us.

Governor Tom Ridge, a U.S. political figure who served as a member of the United States House of Representatives (1983–1995), governor of Pennsylvania (1995–2001), assistant to the president for homeland security (2001–2003), and the first United States secretary of homeland security (2003–2005), got it right when he stated: "You may say Homeland Security is a Y2K problem that doesn't end Jan. 1 of any given year" (Henry, 2002). Homeland security is an ongoing problem that must be dealt with 24/7. Simply, there is no magic on-off switch that we can use to turn off the threat of terrorism in the United States or elsewhere. Consequently, there is no way we can turn off the security switch either. We must stay awake and alert to the threat.

The threat to our security is not only ongoing but is also universal, including potential and real threats from within—from our own citizens (homegrown terrorism). Consider the American Timothy McVeigh, for example, who blew up the government building in Oklahoma City in 1995, killing almost two hundred people, including several children. McVeigh, who bombed the building in revenge for the FBI's Waco, Texas, raid, thought the army (he was a former decorated U.S. Army veteran) had implanted a chip in his butt to track his movements, according to reports.

It is interesting to note that McVeigh, who was no doubt suffering from some type of severe disturbance, acted primarily alone. Actually, McVeigh is the exception that proves the rule—most terrorist acts on America are planned by a group beforehand.

Before 9/11, terrorism was not previously seen as a significant threat to the U.S. However, the fact is that while the U.S. has not experienced continuous 9/11-type terror events in the past, the current climate suggests that the U.S. is at significant risk of further terror activity in the future. Forecasting the future is impossible, but we can look at the past and learn from it. Table 1.1 summarizes past acts of sabotage/terrorism carried out in the U.S. or against U.S. citizens overseas.

Table 1.1. Terror Attack Summaries (within U.S. or against Americans abroad)

Date/Location	*Summary of Attacks*
Sept. 16, 1920, New York City	Bomb exploded in New York City's Wall Street area, killing forty and injuring hundreds. Perpetrators fled country and were never apprehended.
1951–1956, New York City	Between 1951 and 1956, former Consolidated Edison employee set off series of bombs at New York City landmarks, including Grand Central Station and Radio City Music Hall. No deaths. Bomber was later caught and committed to state mental institution.
Jan. 24, 1975, New York City	Bomb set off in historic Fraunces Tavern killed four and injured more than fifty people. Puerto Rican nationals group (FALN) claimed responsibility, and police tied thirteen other bombings to the group.
Nov. 4, 1979, Tehran, Iran	Iranian radical students seized the U.S. embassy, taking sixty-six hostages. Fourteen were released. The remaining fifty-two were freed after 444 days on the day of President Reagan's inauguration.
1982–1991, Lebanon	Thirty U.S. and other Western hostages kidnapped in Lebanon by Hezbollah. Some were killed, some died in captivity, and some were eventually released. Terry Anderson was held for 2,454 days.
April 18, 1983, Lebanon	U.S. embassy destroyed in suicide car-bomb attack; sixty-three dead, including seventeen Americans. The Islamic Jihad claimed responsibility.
Oct. 23, 1983, Lebanon	Shiite suicide bombers exploded truck near U.S. military barracks at Beirut airport, killing 241 marines. Minutes later a second bomb killed fifty-eight French paratroopers in their barracks in West Beirut.
Dec. 12, 1983, Kuwait City, Kuwait	Shiite truck bombers attacked the U.S. embassy and other targets, killing five and injuring eighty.
Sept. 20, 1984, East Beirut, Lebanon	Truck bomb exploded outside the U.S. embassy annex, killing fourteen, including two U.S. military.
Dec. 3, 1984, Beirut, Lebanon	Kuwait Airways Flight 221, from Kuwait to Pakistan, hijacked and diverted to Tehran. Two Americans killed.
April 12, 1985, Madrid, Spain	Bombing at restaurant frequented by U.S. soldiers, killed eighteen Spaniards and injured eighty-two.
June 14, 1985, Beirut, Lebanon	TWA Flight 847 en route from Athens to Rome hijacked to Beirut by Hezbollah terrorists and held for seventeen days. A U.S. Navy diver executed.
Oct. 7, 1985, Mediterranean Sea	Gunmen attacked Italian cruise ship *Achille Lauro*. One U.S. tourist killed. Hijacking linked to Libya.
Dec. 18, 1985, Rome, Italy, and Vienna, Austria	Airports in Rome and Vienna were bombed, killing twenty people, five of whom were Americans. Bombing linked to Libya.
April 2, 1986, Athens, Greece	Bomb exploded aboard TWA Flight 840 en route from Rome to Athens, killing four Americans and injuring nine.

Date/Location	*Summary of Attacks*
April 5, 1986, West Berlin, Germany	Libyans bombed a disco frequented by U.S. servicemen, killing two and injuring hundreds.
Dec. 21, 1988, Lockerbie, Scotland	N.Y.-bound Pan-Am Boeing 747 exploded in flight from a terrorist bomb and crashed into Scottish village, killing all 259 aboard and eleven on ground.
Feb. 26, 1993, New York City	World Trade Center bombing in New York City. Six deaths and one thousand injuries. Intended plan of total structural collapse did not occur. Several members of Middle Eastern extremist organizations convicted of roles in the bombing.
Dec. 7, 1993, New York City	Immigrant opened fire on commuters aboard Long Island Railroad commuter train, killing six and wounding nineteen. Gunman convicted of shootings that were dubbed "the Long Island Railroad Massacre."
April 19, 1995, Oklahoma City	Car bomb exploded outside federal office building, collapsing walls and floors. 168 people were killed, including nineteen children and one person who died in rescue effort. Over 220 buildings sustained damage. Timothy McVeigh and Terry Nichols later convicted in the antigovernment plot to avenge the Branch Davidian standoff in Waco, Tex., exactly two years earlier.
Nov. 13, 1995, Riyadh, Saudi Arabia	Car bomb exploded at U.S. military headquarters, killing five U.S. military servicemen.
Oct. 9, 1995, Hyder, Arizona	Twelve-car Amtrak passenger train traveling from New Orleans to Los Angeles carrying 248 people derailed, killing one and seriously injuring twelve. Investigations indicated that tracks had been deliberately tampered with. Incident classified as deliberate act of sabotage. Ongoing investigation in progress.
June 25, 1996, Dhahran, Saudi Arabia	Truck bomb exploded outside Khobar Towers military complex, killing nineteen American servicemen and injuring hundreds of others. Thirteen Saudis and a Lebanese, all alleged members of Islamic militant group Hezbollah, were indicted on charges relating to the attack in June 2001.
July 27, 1996, Atlanta, Georgia	Bomb exploded during concert at Summer Olympics, killing one and injuring 118.
October 1997, San Francisco, California	An outage at electrical substation in San Francisco affected 126,000 customers. Pacific Gas & Electric said at the time that someone might have deliberately manipulated equipment at the substation to break circuits.
Aug. 7, 1998, Nairobi, Kenya, and Dar es Salaam, Tanzania	Truck bombs exploded almost simultaneously near two U.S. embassies, killing 224 (213 in Kenya and eleven in Tanzania) and injuring about four thousand five hundred. Four men connected with al Qaeda, two of whom had received training at al Qaeda camps inside Afghanistan, were convicted of the killings in May 2001 and later sentenced to life in prison. A federal grand jury had

(*continued*)

Table 1.1. Terror Attack Summaries (within U.S. or against Americans abroad)

Date/Location	*Summary of Attacks*
	Indicted twenty-two men in connection with the attacks, including Saudi dissident Osama bin Laden, who remained at large.
May 2000, Worldwide impact	Computer hacker in Philippines unleashed "Love Bug" computer virus, overloading corporate and government e-mail systems in many countries. Estimated $10–15 billion in damage. Charges against hacker eventually dropped in Philippines because of inadequate laws against computer hacking.
July 25, 2000, Detroit, Michigan	Saboteur with knowledge of Detroit lighting system ripped wiring out of one hundred streetlights in downtown Detroit, leaving live wires exposed and forcing city officials to shut off power to more than six hundred downtown street lights to ensure public safety. Damages estimated at $26,000. One person was arrested and charged.
September 11, 2001, New York City, Washington DC, and Pennsylvania	Terrorist cells of Middle Eastern origination hijacked four commercial jet airliners. Two aircraft used against World Trade Center Towers in New York City, third against Pentagon. Fourth aircraft brought down by passenger intervention. Approximately three thousand killed.
September 2001, U.S.	U.S. Postal Service used as delivery vehicle for anthrax spores contained in letters. Numerous news organizations and government agencies affected across U.S. and in some overseas locations with strong U.S. affiliations. Several killed.
October 7, 2001, Alaska	Intoxicated man shoots at the Alaska pipeline, which results in a seven thousand barrel crude oil spill and the pipeline being shut for several days.

Note: Adapted from Infoplease, "Terrorist Attacks," www.infoplease.com/ipa/A0001454.html.

Case Study 1.1. When Whiskey Sours

My name is W. W. Williams III . . . just call me Willy or Junior. I come from a long line of the same kind, patriots all, thank you very much! Normally, I would not be having this conversation with you . . . no, sir, I wouldn't. Like my father . . . you know, the one who blew up that sugar factory . . . I've never been about talking or jiving or other such dribble. No, sir! Talk is cheap and action is expensive . . . my kind of action, anyway. Besides, you may have heard this introductory incantation or salutation (not sure which word to use here . . . in my genius, I do become confused; don't we all?) before . . . from my deceased father, the one I mentioned earlier; a true patriot, who unfortunately served a very short life sentence; he died a despicable death in a federal prison. So there went another good man, a true patriot, down the proverbial drain, so to speak.

Well, in my family we keep it short, pointed, and irrefutable . . . so let's cut to the chase. Not that I have anything else more pressing to do. No, sir. It's just that I am not used to being out of my psychic ether mode . . . from which I perpetually surfed . . . until I got caught, that is.

Now, at the present time, I am disgustingly sober and in total control of my wits and sanity. But I am not free. No, sir. I sit here on my cement bed in Supermax, Colorado . . . the ultimate prison for the world's ultimate prisoners. Yes, I am one of the worst . . . really bad; bad to the bone marrow, but, alas, sickeningly sober.

Speaking of soberness, I am, like the rest of my family, one of the Pied Piper's (Hunter S. Thompson) followers. It was Thompson who described the good life . . . the psychic Ethernet at its best! In the fog of soberness, I have memorized Thompson's famous description of his driving orgy through Las Vegas—my family's credo.

> In the trunk, we "stow two bags of grass, seventy-five pellets of mescaline, five sheets of high-powered blotter acid, a salt shaker half full of cocaine and a whole galaxy of multicolored uppers, downers, screamers, laughers. . . . A quart of tequila, a quart of rum, a case of Budweiser, a pint of raw ether and two dozen amyls."

As my daddy always said: Doesn't it just torque your jaws that some of us know how to live . . . and the rest just exist. Geez, what a man . . . Thompson is my all-time hero . . . for sure.

Sigh! I guess we should get back to the topic and conversation at hand. I was given this chance to state my case . . . to explain why I put my gang together to attack the U.S. electrical grid and put it out of commission and kill a few of those bleeding heart liberals (just call them bleeders) in the bargain. So let's get to it. Let me tell you my story. It ain't pretty in some folks' way of thinking, I suppose. But that . . . others' way of thinking has never bothered me too much . . . No, sir. I do it . . . and did it my way. And I would be happy forever sitting on this cement bed speaking to you all if they would just let me have my ether. Oh well, what the hell . . . can't have it all . . . not anymore, anyway.

I joined Greenway Coalition in 1970 during one of the first Earth Day celebrations in Ohio. I was eleven years old at the time. During my storied career with Greenway I was involved in some glorious undertakings. For example, I was with the team that tree spiked several old growth trees on the Olympic Peninsula, Washington State. In my last incident there, I personally drove several metal rods into various tree trunks and saved the trees. Unfortunately, a couple of loggers died and others were injured when their saws struck the spikes and the saws kicked back and decapitated two loggers and seriously wounded the others . . . just collateral damage . . . necessary collateral damage, in my opinion.

Well, as it turned out spiking all those trees was great but looking at all the blood and guts later was a bit overwhelming, so I shifted gears from tree spiking to monkey wrenching or ecotaging.

My monkey-wrenching activities also dealt with saving old growth forests. However, instead of spiking trees I ecotaged logging equipment. This turned out to be an ideal practice because I was able to prevent a lot of tree destruction because the logging equipment all malfunctioned and in some cases massively destroyed itself when they tried to start it. Also, only a few loggers were killed or injured by the monkey-wrenched equipment . . . not that I cared, you understand . . . so I guess what it turned out to be was a win-win situation . . . for a while, at least.

Over the years Greenway Coalition grew in numbers and sophistication . . . and I grew right along with it. In the late nineties, we shifted gears from forestry to other areas of interest, including committing arson to preserve the pristine wilderness. The Vail Mountain arson attack, for example, where we torched buildings that were going up . . . those buildings would have destroyed an outdoor paradise. Man, there are just some places on earth where man should not intrude. Later we returned to Oregon and torched the main office of a logging company. In 1999 we even went to New Zealand to uproot fields of genetically engineered potatoes. The engineering of foodstuffs should be left to Mother Nature alone and definitely not to man. A few years later we burnt down a two-hundred-unit condominium being built in southern California. We put out the word that "if you build it, we will trash it." Then again we shifted gears. We understood that one of the biggest mistakes mankind had made was in developing the automobile. These horrendous machines are nothing but polluters of the environment, with low fuel efficiency and destroyers of things they happen to be run into. So we decided to attack several car dealerships, garages, storehouses in Los Angeles. Most of the autos we attacked were those gas-guzzling, road-hogging SUVs or Hummers.

A few years ago we decided to change our focus again. We also decided that arson was a bit too damaging . . . we would take out the bad guys for sure, but in some cases we actually burned down the surrounding forest or city blocks, etc. Collateral damage for sure, but we decided that we had made our point in that specific area. The fact is we had put a lot of loggers and car dealerships out of business and the spotted owl sleeps better at night in cleaner air because of our actions.

One of the things that really surprised us, devastated us (if you really must know) was the distorted press and other news media venues out there in la-la land that had the nerve to address these heroic actions of ours as acts of vandalism and other forms of destruction. Some even had the audacity to label our actions terrorism or the acts of terrorists. Can you believe that? One thing is certain, the new homeland security secretary, Janet Napolitano, would not be pleased with anyone, including me, being labeled a "terrorist."

So here I vegetate in this cement block, locked up like one of those so-called terrorists (no offense intended to the new secretary of homeland security). One media outlet had the audacity to call me crazy . . . a real nut case . . . a crank! Can you believe that, man!

Well, speaking of crank, it (the crank) is partially to blame (I guess) for me being entombed in concrete. Why? Well, guess I ought to explain the rest of this fiasco so you will understand my plight . . . my situation, my travesty, so to speak.

It came to me one stormy night like a bolt of lightning . . . illuminating the way for me and my followers. You see, because of my innate ability to plan and to kill and to destroy I became leader of my own so-called fringe group to be. We had no name or title . . . no, sir. We had no need for any of that . . . interference with and destruction of man's intrusion on nature was our goal and our mantra.

Back to that stormy night.

I was tooting up some crack and crank, chased by several shots of my favorite straight whiskey . . . I always drink my whiskey straight . . . none of that mix to make it sour for me. Anyway, we were camping out in a tent in the Texas panhandle one night. Now, when I say we, I mean me and a few of my closest associates . . . you know, the ones I could trust the most, women, of course . . . actually they were the ones with the dope and whiskey. So, we sat inside the tent combining a bit of crack and crank into smokes and chasing it all with America's finest whiskey when a tremendous storm came up and things got loud, windy, and very wet.

After several toots and down the hatches, so to speak, I decided to step outside into the wind, downpour and streaking lightning . . . I shook my fist at the heavens and demanded that someone up there bring this mess to a halt . . . simply, enough was enough. And that is when it hit me!

No, not the lightning . . . though I could have used a battery recharge about then . . . no, sir, it was not the lightning that hit me. It was . . . it was . . . gee, how do I describe it? As I said, the wind was howling, sky lit up with streaks of lightning, thunder pounding, and rain everywhere. But what got my attention . . . made me almost sober . . . was those long, ghastly, sharpened fingers slashing, whipping, lashing, whirling viciously though the air above me and off to the west as far as I could see in the intermittent light. What a scene. When is enough enough, man? Of course, in time I came to recognize those arms severing the air for what they were . . . man-made (Janet Napolitano would be proud) giant scissor-handed hulking grotesque monsters slashing and groping their way across the high plains . . . wind turbos from hell!

Now I am sure one of you literary geniuses reading this account has equated my experience to that of tilting (jousting) at windmills, Don Quixote fashion . . . nothing could be further from the truth . . . I admit to having been somewhat stoned during my adventure, but crazy like Don Quixote? No. No way. Hell, no! I can assure you that I do not have a crazy bone in my body. None!

For years after the camping experience in one of America's largest wind farms, I have not been able to get those horrendous, grotesque mechanical images out of my mind. Alas, and it was that one particular spectral experience that shaped the rest of my life and landed me here to vegetate in this cement garden.

After that night of stuporous manifestations from hell (there is no other way to describe it, thank you very much), I came to the realization—my Eureka moment—that I had finally found my mission in life . . . that is, to clear the land of wind turbines. I was so excited!

Right about now, you might be asking yourself what is the problem with wind turbines? Aren't wind turbines ecologically more practical than fossil-fueled energy producers? Is it not true that wind turbines do not leave a carbon footprint like other sources of fuel? Most importantly, is wind power renewable?

Sure. All that is true . . . but think about the land area those wind farms take up. Think about the noise they produce. Think about the grotesque-looking image they make against the usually otherwise panoramic landscapes. Then there are those birds and bats that those turbine blades hack to pieces. Now understand, I am not a fan of birds and bats . . . to me they are nothing more than direct descendants of dinosaurs; they make good fossils, etc. . . . and that is about all. Besides, all those avian would-be monsters do is eat and poop . . . mostly poop. Who needs all that? On the other hand, birds and bats have as much right to exist as we do . . . dinosaur-like or not, they are living things. Thus, they must be protected. Protecting them became my ultimate mission in life.

So during the few intervening years before my ultimate coup de grâce, I torched a few SUV dealerships here and there and also planned out my assault on those terrible wind turbine farms. However, being the genius I am it was clear to me from the beginning of my planning stage that I could not attack thousands of wind turbines alone. No way, José!

I had always worked lone-wolf fashion. Working alone, outside of any command structure, just works better for me. I am and always have been a part of the Greenway Coalition, of course, but as an appendage only—I'm connected by deeds, actions, ideological and philosophical identification, but not by any direct group communication tie-in. I have found that my tactics and methods, conceived by me alone, are best directed and controlled by myself. I consider myself in a league all by myself and certainly far above my heroes Timothy McVeigh, Theodore Kaczynski (the Unabomber), and Eric Robert Rudolf of 1996 Atlantic Olympics fame. These three all epitomize my lone-wolf characteristics. The difference between us? They got caught and I never intended to follow suit . . . but you know what they say about the best-laid plans of both . . . etc.

Anyway, this time around . . . on my wind turbine destruction caper, I knew that it would take a group of dedicated individuals (like myself, of course) to pull this one off. However, like I mentioned earlier, I like my whiskey straight and do not sour it by mixing; I do not like mixing with people . . . with a group. Too many nuts and bolts and therefore too many chances for something to go wrong . . . sours the operation. Each additional person added to the team is just another weak link, so to speak.

Then I had another revelation. Attacking individual wind turbines made little sense and was basically impractical. Attacking entire wind farms was also a problem because there was no way we could muster up enough explosives and correctly place them undetected to do the damage I wanted done. So I thought through the matter and determined that the best bang for the buck could be obtained by attacking the electrical substations that were fed electrical power by wind turbines. Actually, such an attack by a group of activists had already occurred in 2005 in South Middleton, Pennsylvania. The activists shot out a transformer at the power substation, which knocked out electrical power for three counties and twenty thousand customers.

I recruited a dedicated team of forty men and women from all parts of the U.S. All these folks were old hands at ecotage and all were members of Greenway Coalition . . . more than eager to participate in any kind of destruction of U.S. infrastructure . . . my kind of people. I decided that we would meet in Sturgis, South Dakota . . . in August of that year, during the Sturgis Rally . . . you know, the annual gathering of more than four hundred thousand bikers for a week of fun and disobedience on two wheels, usually resulting in around 450 individual arrests for one crime or another . . . don't you just love 'em, man!

Sturgis seemed the obvious choice. I told all my team members to show up in blue jeans, leather jackets, unshaven, unkempt, and just plain ready to fit in with the biker crowd. And they did, and we did . . . we fit in . . . no one seemed to notice we even existed . . . which was what I wanted . . . just flies on the wall, preparing for the ultimate egg-laying attack.

My team met twice. We met in a picnic area just outside of Sturgis along I-90 . . . perfect location. The forty-member team was briefed on the plan. The plan? Well, it was rather easy, definitely not complicated . . . I was working with a team of true patriots, but many were a bit short on brain power, a few were short a few microchips, and a few had very little common sense . . . good killer types but not too bright, if you get my drift. Each team member was to work alone in a designated location in an area where wind turbines fed electricity to electrical substations, which in turn fed electricity to a town or city or major distribution network. All of these locations were in California, Arizona, New Mexico, Texas, and West Virginia. My plan was to put the lights out in as many places as possible and to silence those horrible air-cutting wind turbines. Now, I knew the destruction would be temporary, but I also knew that we could and would strike again and again . . . each time making reconstruction just a bit too expensive in the long run . . . thus, I would win!

Each team member was outfitted with a 30.06 rifle. We used a couple of hours in the Sturgis picnic area for some target practice. I outfitted each of the forty shooters with 180-grain bullets. I knew we needed the high speed, high energy impact of those 180-grain bullets to do the damage required to put those substation transformers out of commission. Turned out that the shooters were pretty good at it . . . and it was funny, when you have forty people shooting off high-powered rifles, you would think someone would notice. Actually, people did notice . . . they just smiled and waved to us as they passed by. I guess in that part of the country all you need is a horse, a Harley, a meth lab, or a gun to fit right in . . . fitting right in is what we are/were all about.

I planned the attack for Labor Day. I did this on purpose, of course, because number one it was a holiday . . . when those so-called working folks could care less about anything. Also, symbolically Labor Day was perfect . . . a sort of twisted protest on my part . . . since it is my belief that the commie labor unions and their constant chant of "give me more, more, and more and we will give you less, less, and less in return" is what has destroyed this great country of ours . . . really torques my jaws. You think I am exaggerating? Take a look at the American auto manufacturing folks . . . how well are they doing? Why has almost all other manufacturing left this country and moved off to commie-land? Why are so many red-blooded Americans unemployed? Unions . . . it is all about unions. They have destroyed the land I love. Well, now it is get even time, my friends . . . the worm has turned.

Anyway, along with target practice with the rifles, I needed to brief each shooter on where to aim the rounds at an electrical substation for greatest effect. In my cement heaven, because my case is still undergoing another appeal, I am allowed to keep certain legal documents in my cell. So if you look at the photo of the substation, you will see the same photo we used for training.

Well, you might be wondering why I freely admit all that happened, even showing you the substation training photo, especially when my case is under appeal. My appeal is based on police harassment, failure to be properly Mirandized, and also jailor brutality . . . you know how those brutal jailhouse pigs act, those same want-to-be pigs who can't cut the mustard as real cops, they are bullies . . . each and every one. Unfortunately, this is my fourth appeal . . . so chances are slim I will win,

Electrical substation

anyway . . . though some of that ether right now would be nice, thank you very much!

Anyway, let's get back to the fiasco . . . and a fiasco it turned out to be . . . indeed. As I mentioned, I had set the date for the simultaneous attack on forty different major city electrical substations for Labor Day. Now these attacks were to take place in different time zones, so I had each person set their watches to the same time and planned the attack time at 10:00 p.m. No problem there. The problem arose at a huge power substation outside Amarillo, Texas. Unfortunately, as it turned out the woman I assigned to this station, JoAnn Garner-Potts II, was the ultimate ding-a-ling-dong and meth head . . . an hour before the 10:00 p.m. attack time she was observed wobbling around with rifle in hand in front of the electrical substation, her target, scoping the place out. Well, needless to say someone called the pigs and they showed up in force and threw her methed-up body to the ground, brutalized the hell out of her, etc. etc. etc. Within five minutes, JoAnn Garner-Potts II told her capturers everything . . . I mean everything. Before any of the rest of us, thirty-nine of us on station, could do our damage, we were all arrested, brutalized by the pigs, incarcerated by the bleeders that run the courts, and here I sit in this cement palace . . . surrounded by some of the worst killers known to modern man. What a waste! And that old axiom keeps ringing in my ears: all I want to know is where I am going to die so I will never go there—my problem? I am there—for life and death.

So now it is two years later, and here I sit after one of the longest, most ridiculous trials in criminal history . . . topped off by three losing appeals and, as mentioned, waiting on the outcome of another. Everyone was out to get me. I hadn't killed or injured anyone at the substations. I just don't get it!

Sitting here now and basking in the total disgrace of failure, only three things bother me: One, I got caught by totally incompetent, dysfunctional dead heads; two, during the trial and all the publicity that followed, they classified me as an American terrorist (Janet Napolitano would not be pleased). "Terrorist?" No way. It is true I have killed . . . collateral damage . . . but the overall goal of saving the environment is all that mattered . . . the rest is poppycock . . . trivial to the bone. I am not a terrorist. I really don't think the press, law, legal authorities, or the people know what a terrorist really is. Maybe someone ought to write a book on the subject to inform everyone what a terrorist really is. One thing is certain; no, sir, I am not a terrorist. Finally, the third item I am upset about is that I did not stick to my own mantra. That is, as I mentioned earlier, when you add mix to straight whiskey you turn it sour. Likewise, when you trade lone-wolf operations for mixed ops, you sour the operation.

Keep in mind that many define "free society" in different ways. This is important to us in this text because in our view, in order to be truly safe from the terrorist threat (if that is possible, and we have our doubts) we must give up certain freedoms and accept closer scrutiny and vigilance in regard to actions that we normally assume to be no one's business but our own. There are differing opinions on this topic, and we fully accept that. Again, we each define freedom in our own way—and this is our right as free people.

Precise definition of key terms is required for us to clearly understand the purview of this text. Keep in mind that the key terms and their definitions presented below may differ from those used in the U.S. Department of Homeland Security National Infrastructure Protection Plan.

WHAT IS TERRORISM?

Since 9/11, we have heard it said by many of our students (and others) that there is controversy about the definition of the politically charged word *terrorism.* Terrorism, like pollution, is a judgment call. For example, if two neighbors live next door to an air polluting facility, one neighbor who has no personal connection with the polluting plant is likely to label the plant's output as pollution. The other neighbor, who is an employee of the plant, may see the plant's pollution as dollar bills—dollars that are his livelihood. On the terrorism front, when someone deliberately spikes a tree to prevent loggers from cutting it down, the tree spiker might feel he is a patriot and not a terrorist. On the other hand, the logger who has to take the tree down and puts life and limb at risk in taking down the spiked tree has little doubt in his mind on what to call the tree spiker, and it certainly has nothing to do with patriotism. Thus, what we are saying here is that along with defining pollution, defining terrorism may be a judgment call, especially in the view of the terrorists.

To make our point on the countless differing views in defining terrorism, consider, for example, that if we were to ask one hundred different individuals to define terrorism, we would likely receive a hundred different definitions. As a case in point, consider the following: if we were to ask a hundred different individuals to describe the actions of Willy in case study 1.1, how would they describe him and his actions? You might be surprised—we were. In 2000 and again in 2007, pre- and post-9/11, after reading Willy's wind turbine and electrical power station incident, a hundred randomly selected Old Dominion University juniors and seniors studying environmental health ("Generation Y" students ranging in age from twenty to forty years old) were asked to reply to a nonscientific survey questionnaire. The two questions and the students' responses to this unscientific survey are listed in table 1.2.

From the Old Dominion University nonscientific survey it is clear that the students' perceptions of Willy's actions in case study 1.1 shifted dramatically from pre-9/11 to post-9/11. For example, when asked to select the best pre-9/11 descriptor to describe Willy, "crazy" and "insane" ranked high; however, after 9/11, the students' perception shifted away from "crazy" and "insane" to "terrorist." Likewise, the students' pre-9/11

Table 1.2. Student Responses, Pre- and Post-9/11

	Number of Responses	
*Students' Responses Descriptors**	*Pre-9/11 (2000)*	*Post-9/11 (2007)*
Question 1: In your opinion, Willy was		
crazy	30	0
a disgruntled former employee	10	0
insane	14	5
misguided	1	0
a cold-blooded murderer	14	16
a misfit	1	0
deranged	6	4
a lunatic	5	1
a bully	13	3
a terrorist	1	63
not sure	0	0
Totals	100	100
Question 2: In your opinion, Willy's actions are best described as		
madness	55	1
frustration	9	2
desperation	4	0
dysfunctional thinking	2	0
legitimate concern	1	0
threatening	6	1
terrorism	5	84
workplace violence	0	0
not sure	18	12
Totals	100	100

* Student response descriptors were provided to the students by the instructors.

responses on Willy's actions ranked high in "madness"; his actions post-9/11 overwhelmingly were described as "terrorism."

It is interesting to note that the 2000 year group reported prior to the September/October 2001 anthrax attacks and prior to 9/11; the student group's responses were provided after such events as the World Trade Center attack of 1993 and Timothy McVeigh's 1995 mass murder of the occupants of the federal building in Oklahoma City, Oklahoma. This may help to explain why the year 2000 students were somewhat reluctant to describe Willy's actions as terrorism and/or to label him as a terrorist.

Terrorism by Any Other Name Is . . .

From the preceding discussion we might want to buy into the argument that terrorism is relative, a personal judgment. But is it really relative? And if it is relative . . . relative to what? Is it a personal judgment? What is terrorism? Take your choice. Seemingly, there is an endless list of definitions and no universally accepted definition of terrorism, the main theme of this text. Let's review a few of these definitions.

Standard Dictionary Definition of Terrorism

After reviewing several dictionaries, we have found this fairly standard definition of terrorism:

> The unlawful use or threatened use of force or violence by a person or an organized group against people or property with the intention of intimidating or coercing societies or governments, often for ideological or political reasons.

America's *National Strategy for Homeland Security* defines terrorism as follows:

> Any premeditated, unlawful act dangerous to human life or public welfare that is intended to intimidate or coerce civilian populations or governments (NSHS 2006).

The U.S. State Department defines terrorism thus:

> Premeditated, politically motivated violence perpetrated against noncombatant targets by subnational groups or clandestine agents (U.S. Congress 2005).

The FBI definition of terrorism is as follows:

> The unlawful use of force or violence against persons or property to intimidate or coerce a Government, the civilian population, or any segment thereof, in furtherance of political or social objectives (FBI 2006).

Note that the FBI divides terrorism into two categories: domestic (homegrown), involving groups operating in and targeting the U.S. without foreign direction, and international, involving groups that operate across international borders and/or have foreign connections.

Well, at this point the obvious question is: Do you now know what terrorism is? That is, can you definitely define it? If you can't define it, you are not alone—not even the U.S. government can definitively define it. Maybe we need to look at other sources—views from the real experts on terrorism.

Osama bin Laden's View on Terrorism

> Wherever we look, we find the U.S. as the leader of terrorism and crime in the world. The U.S. does not consider it a terrorist act to throw atomic bombs at nations thousands of miles away [Japan during World War II], when those bombs would hit more than just military targets. Those bombs rather were thrown at entire nations, including women, children, and elderly people. . . . (Bergen 2002, 21–22).

Another View

This view is from court testimony on terrorism from Ramzi Ahmed Yousef, who helped organize the first terrorist attack on the World Trade Center:

> You keep talking also about collective punishment and killing innocent people to force governments to change their policies; you call this terrorism when someone would kill innocent people or civilians in order to force the government to change its policies. Well, when you were the first one who invented this. . . .
>
> You were the first one who killed innocent people, and you are the first one who introduced this type of terrorism to the history of mankind when you dropped an atomic bomb which killed tens of thousands of women and children in Japan and when you killed over a hundred thousand people, most of them civilians, in Tokyo with fire bombings.
>
> You killed them by burning them to death. And you killed civilians in Vietnam with chemicals as with the so-called Orange agent. You killed civilians and innocent people, not soldiers, innocent people every single war you went. You went to wars more than any other country in this century, and then you have the nerve to talk about killing innocent people.
>
> And now you have invented new ways to kill innocent people. You have so-called economic embargo which kills nobody other than children and elderly people, and which other than Iraq you have been placing the economic embargo on Cuba and other countries for over 35 years. . . .
>
> The government in its summations and opening said that I was a terrorist. Yes, I am a terrorist and I am proud of it. And I support terrorism so long as it was against the United States Government and Israel, because you are more than terrorists; you are the one who invented terrorism and using it every day. You are butchers, liars and hypocrites (excerpts from statements in court 1998).

And, finally, here is an old cliché on a terrorist:

> One man's terrorist is another man's freedom fighter.

Again, from the preceding points of view, it can be seen that defining terrorism or the terrorist is not straightforward and never easy. Even the standard dictionary definition leaves us with the vagaries and ambiguities of other words typically associated with terrorism, such as in the definitions of *unlawful* and *public welfare* (Sauter and Carafano 2005).

Raphael Perl (2004) in a Congressional Research Service (CRS) report points out that one definition widely used in U.S. government circles and incorporated into law defines *international terrorism* as terrorism involving the citizens or property of more than one country.

Terrorism is broadly defined as politically motivated violence perpetrated against noncombatant targets by subnational groups or clandestine agents. For example, kidnapping of U.S. birdwatchers or bombing of U.S.-owned oil pipelines by leftist guerrillas in Colombia would qualify as international terrorism. In 22 U.S.C. 2656f, a *terrorist group* is defined as a group which practices terrorism or has significant subgroups that practice terrorism. Perl (2004) points out that one of the shortfalls of this traditional definition is its focus on groups and its exclusion of individual ("lone-wolf") terrorist activity, which has recently risen in frequency and visibility.

At this point, readers may wonder, "Why should we care—that is, what difference does it make what the definition of terrorist or terrorism is?" Definitions are important because in order to prepare for the terrorism contingency, domestic or international, we must have some feel, as with any other problem, for what it is we are dealing with. We are fighting a war of ideas. We must attempt to understand both sides of the argument, even though the terrorist's side makes no sense to an American or other freedom-loving occupant of the globe.

The *Washington Times* (Napolitano Tells It Like It Isn't 2009) reports that in an interview with German magazine *Der Spiegel*, the secretary of homeland security said that she's rejected the word *terrorism* and has instead renamed terrorist acts using the euphemism "man-caused disasters," because "it demonstrates that we want to move away from the politics of fear, toward a policy of being prepared for all risks that can occur." This is analogous to stating that we should refer to serial killings, serial rapes, and serial arson as man-caused afflictions. After all, we do not want to create fear about serial offenders. Get real! The authors of this text suggest that Secretary Napolitano read the remarks we mentioned earlier by Ramzi Ahmed Yousef:

> Yes, I am a terrorist and I am proud of it. And I support terrorism as long as it was against the United States Government and Israel. . . .

Finally, while it is difficult to pinpoint an exact definition of terrorism (but not difficult to rename it and call it something else), we certainly have little difficulty in identifying it when we see it, when we feel it, when we suffer from it. Consider, for example, in the earlier account of Willy's actions in case study 1.1 and the various ecotage capers he was involved in. Put yourself in the place of those workers who were harmed or killed by his actions, people who were simply working to support themselves and their families. When the events occurred, none of the victims could have known that an American terrorist had caused the act of terrorism on U.S. soil that killed them or their coworkers and/or badly maimed and disfigured others. No, they did not know that. However, if not killed instantly, there is one thing they knew for certain; they knew that crushing feeling of terror as they struggled to breathe, to recover, to survive. By any other name terrorism is best summed up by an absolute feeling of terror—nothing judgmental about that—just terror with a capital *T*.

A VOCABULARY OF HATE

In America, there are plenty of hate groups that claim peace and brotherhood, but when their actions are responsible for death and destruction, they are identified for what they really are. (Lindsey 2001)

After 9/11 several authors published and the media transmitted seemingly endless accounts of various hate groups operating throughout the globe. Overnight Americans became aware of various theories, philosophies, and terminology very few had ever heard of or thought about. This trend is ongoing—never ending.

Various pundits, so-called experts on the "new" genre of terrorism, have stated that for Americans to understand why foreign terrorists behead innocent people (or anyone else, for that matter) on television or blow up hospitals full of the sick or wounded or schoolhouses full of children, they must get inside the mind of a terrorist—enter the confines of the black box.

The average American might ask: "Get inside the mind of a terrorist? How the hell do you get inside the mind of madmen?"

This is where we make our first mistake, thinking the terrorists act in the manner they do because they are mad, nonrational, disturbed, or psychotic. In the case of Timothy McVeigh, we might be able to characterize him and his actions in this manner. Yet McVeigh, as mentioned earlier, is the exception that proves the rule—terrorists' attacks, by real terrorists, are primarily planned beforehand by a group. It is important to remember that McVeigh acted primarily alone.

The terrorists that crashed aircraft into the Twin Towers, Pentagon, and that farm field in Pennsylvania were all of the same mindset; they worked as a group. Likewise, the terrorists that attack Baghdad, Afghanistan, and Pakistan every day work as a group. Terrorists that did all the damage in Bali and Spain and elsewhere acted as a group. Thus, though we would like to classify all the terrorists as we classify Timothy McVeigh, we can't do that. One madman working alone is something we can reasonably assume. However, thinking that hundreds or thousands of like-minded madmen all work in groups is a stretch—even though it is true. The cold-blooded manner in which terrorists go about their business suggests that they are not crazy, insane, or mad, but instead extremely harsh and calculating. If we dismiss them as madmen, we underestimate their intelligence. When we do that, we lose. No, we cannot underestimate the enemy—the terrorists. They are smart, cold-blooded, and calculating. In order to protect our critical energy infrastructure, we must be smarter and expect the unexpected—we must be proactive and not just reactive in implementing our countermeasures. Understanding is important. For example, understand that the Koran does not condone murder and suicide; instead, it provides injunctions against suicide and murder. When the Muslim terrorist commits murder or suicide in the name of Islam and the Koran, what he or she is really doing is changing the meaning of the words *murder/suicide* to mean martyrdom. No, getting into the mind of a terrorist is not the solution. Instead, in regard to terrorists, we must ensure they do not get into our minds. Simply, we must be smarter than the enemy.

The Language of Terrorism

Anyone who is going to work at improving the security of America's critical infrastructure must be well versed in the goals and techniques used by the terrorists. Moreover, we cannot implement effective countermeasures unless we know our vulnerabilities. Along with this, we must not only understand what terrorists are capable of doing but also have some feel for their language or vocabulary, which will help us to understand where they are coming from and where they might be headed, so to speak.

As with any other technical presentation, understanding the information presented is difficult unless a common vocabulary is established. Voltaire said it best: "If you wish to converse with me, please define your terms." It is difficult enough to understand terrorists and terrorism (or man-caused disasters); thus, we must be familiar with terms they use and that are used to describe them, their techniques, and their actions.

Definition of Terms

Abu Sayyaf—Meaning "bearer of the sword"; the smaller of the two Islamist groups whose goal is to establish an Iranian-style Islamic state in Mindanao in the southern Philippines. In 1991, the group split from the Maro National Liberation Front. With ties to numerous Islamic fundamentalist groups, they finance their operations through kidnapping for ransom, extortion, piracy, and other criminal acts. It is also thought that they receive funding from al Qaeda. It is estimated that there are between two hundred and five hundred Abu Sayyaf terrorists, mostly recruited from high schools and colleges.

acid bomb—A crude bomb made by combining muriatic acid with aluminum strips in a two-liter soda bottle.

aerosol—A fine mist or spray, which contains minute particles.

Afghanistan—At the time of 9/11, Afghanistan was governed by the Taliban, and Osama bin Laden called it home. Amid U.S. air strikes, which began on October 7, 2001, the U.S. sent in more than three hundred million dollars in humanitarian aid.

In December 2001, Afghanistan reopened its embassy for the first time in more than twenty years.

aflatoxin—A toxin created by bacteria that grow on stored foods, especially on rice, peanuts, and cottonseeds.

agency—A division of government with a specific function, or a nongovernmental organization (e.g., private contractor, business, etc.) that offers a particular kind of assistance. In the incident command system, agencies are defined as jurisdictional (having statutory responsibility for incident mitigation) or assisting and/or cooperating (providing resources and/or assistance).

airborne—Carried by or through the air.

air marshal—A federal marshal whose purpose is to ride commercial flights dressed in plain clothes and armed to prevent hijackings. Israel's use of air marshals on El Al is credited as the reason Israel has had only a single hijacking in over thirty years. The U.S. started using air marshals after September 11. Despite President Bush's urging, there are not enough air marshals to go around, so many flights do not have them.

al-Gama'a al-Islamiyya (The Islamic Group, IG)—Islamic terrorist group that emerged spontaneously during the 1970s in Egyptian jails and later in Egyptian universities. After President Sadat released most of the Islamic prisoners from prisons in 1971, groups of militants organized themselves in groups and cells, and al-Gama'a al-Islamiyya was one of them.

al Jazeera—Satellite television station based in Qatar and broadcast throughout the Middle East. Al Jazeera has often been called the "CNN of the Arab world."

alpha radiation—The least-penetrating type of nuclear radiation. Not considered dangerous unless particles enter the body.

al Qaeda—Meaning "the base"; an international terrorist group founded in approximately 1989 and dedicated to opposing non-Islamic governments with force and violence. One of the principal goals of al Qaeda was to drive the U.S. armed forces out of the Saudi Arabian peninsula and Somalia by violence. Currently wanted for several terrorist attacks, including those on the U.S. embassy in Kenya and Tanzania as well as the first and second World Trade Center bombings and the attack on the Pentagon.

al Tahwid—A Palestinian group based in London that professes a desire to destroy both Israel and the Jewish people throughout Europe. Eleven al Tahwid were arrested in Germany as they allegedly were about to begin attacking that country.

American Airlines Flight 11—The Boeing 767 carrying eighty-one passengers, nine flight attendants, and two pilots, which was hijacked and crashed into the north tower of the World Trade Center at 8:45 a.m. eastern time on September 11, 2001. Flight 11 was en route to Los Angeles from Boston.

American Airlines Flight 77—The Boeing 757 carrying fifty-eight passengers, four flight attendants, and two pilots, which was hijacked and crashed into the Pentagon at 9:40

a.m. eastern time on September 11, 2001. Flight 77 was en route to Los Angeles from Dulles International Airport in Virginia.

ammonium nitrate-fuel oil (ANFO)—A powerful explosive made by mixing fertilizer and fuel oil. The type of bomb used in the first World Trade Center attack as well as the Oklahoma City bombing.

analyte—The name assigned to a substance or feature that describes it in terms of its molecular composition, taxonomic nomenclature, or other characteristic.

anthrax—An often fatal infectious disease contracted from animals. Anthrax spores have a long survival period, the incubation period is short, and disability is severe, making anthrax a bioweapon of choice by several nations.

antidote—A remedy to counteract the effects of poison.

antigen—A substance that stimulates an immune response by the body's immune system. The body recognizes such substances as foreign and produces antibodies to fight them.

antitoxin—An antibody that neutralizes a biological toxin.

Armed Islamic Group (GIA)—An Algerian Islamic extremist group that aims to overthrow the secular regime in Algeria and replace it with an Islamic state. The GIA began its violent activities in early 1992 after Algiers voided the victory of the largest Islamic party, Islamic Salvation Front (FIS), in the December 1991 elections.

asymmetric threat—The use of crude or low-tech methods to attack a superior or more high-tech enemy.

Axis of Evil—Iran, Iraq, and North Korea, as mentioned by President G. W. Bush during his State of the Union speech in 2002 as nations that were a threat to U.S. security due to harboring terrorism.

Baath Party—The official political party in Iraq until the U.S. debaathified Iraq in May 2003, after a war that lasted a little over a month. Saddam Hussein, the former ruler of the Baath Party, was targeted by American-led coalition forces and fled. Baath Party members have been officially banned from participating in any new government in Iraq.

Beltway Sniper—For nearly a month in October 2002, the Washington D.C., Maryland, and Virginia area was the hunting ground for forty-one-year-old John Allen Muhammad and seventeen-year-old Lee Boyd Malvo. Dubbed "the Beltway Sniper" by the media, they shot people at seemingly random places such as schools, restaurants, and gas stations.

bioaccumulative—Substances that concentrate in living organisms rather than being eliminated through natural processes, as in the breather of contaminated air or those who drink or live in contaminated water or eat contaminated food.

biochemical warfare—Collective term for use of both chemical warfare and biological warfare weapons.

biochemterrorism—Terrorism using as weapons biological or chemical agents.

biological ammunition—Ammunition designed specifically to release a biological agent used as the warhead for biological weapons. Biological ammunition may take many forms, such as a missile warhead or bomb.

biological attack—The deliberate release of germs or other biological substances that cause illness.

Biosafety Level 1—Suitable for work involving well-characterized biological agents not known to consistently cause disease in healthy adult humans, and of minimal potential hazard to lab personnel and the environment. Work is generally conducted on open bench tops using standard microbiological practices.

Biosafety Level 2—Suitable for work involving biological agents of moderate potential hazard to personnel and the environment. Lab personnel should have specific training in handling pathogenic agents and be directed by competent scientists. Access to the lab should be limited when work is being conducted, extreme precautions should be taken with contaminated sharp items, and certain procedures should be conducted in biological safety cabinets or other physical containment equipment if there is a risk of creating infectious aerosols or splashes.

Biosafety Level 3—Suitable for work done with indigenous or exotic biological agents that may cause serious or potentially lethal disease as a result of exposure by inhalation. Lab personnel must have specific training in handling pathogenic and potentially lethal agents and be supervised by competent scientists who are experienced in working with these agents. All procedures involving the manipulation of infectious material are conducted within biological safety cabinets or other physical containment devices, or by personnel wearing appropriate personal protective clothing and equipment. The lab must have special engineering and design features.

Biosafety Level 4—Suitable for work with the most infectious biological agents. Access to the two Biosafety Level 4 labs in the U.S. is highly restricted.

bioterrorism—The use of biological agents in a terrorist operation. Biological toxins would include anthrax, ricin, botulism, the plague, smallpox, and tularemia.

Bioterrorism Act—The Public Health Security and Bioterrorism Preparedness and Response Act of 2002.

biowarfare—The use of biological agents to cause harm to targeted people either directly, by bringing the people into contact with the agents, or indirectly, by infecting other animals and plants, which would in turn cause harm to the people.

blister agents—Agents that cause pain and incapacitation instead of death and might be used to injure many people at once, thereby overloading medical facilities and causing fear in the population. Mustard gas is the best-known blister agent.

blood agents—Agents based on cyanide compounds. More likely to be used for assassination than for terrorism.

botulism—An illness caused by the botulinum toxin, which is exceedingly lethal and quite simple to produce. It takes just a small amount of the toxin to destroy the central nervous system. Botulism may be contracted by the ingestion of contaminated food or through breaks or cuts in the skin. Food supply contamination or aerosol dissemination of the botulinum toxin are the two ways most likely to be used by terrorists.

Bush Doctrine—The policy that holds responsible nations that harbor or support terrorist organizations and says that such countries are considered hostile to the U.S. From Karen Hughes, White House counselor to the president: "A country that harbors terrorists will either deliver the terrorists or share in their fate. . . . People have to choose sides. They are either with the terrorists, or they're with us."

BWC—Officially known as the "Convention on the Prohibition of Development, Production, and Stockpiling of Bacteriological (Biological) and Toxin Weapons and Destruction." The BWC works toward general and complete disarmament, including the prohibition and elimination of all types of weapons of mass destruction.

Camp X-Ray—The Guantanamo Bay, Cuba, detention camp that houses al Qaeda and Taliban prisoners.

carrier—A person or animal that is potentially a source of infection by carrying an infectious agent without visible symptoms of the disease.

cascading event—The occurrence of one event that causes another event.

causative agent—The pathogen, chemical, or other substance that is the cause of disease or death in an individual.

cell—The smallest unit within a guerrilla or terrorist group. A cell generally consists of two to five people dedicated to a terrorist cause. The formation of cells is born of the concept that an apparent "leaderless resistance" makes it hard for counterterrorists to penetrate.

chain of custody—The tracking and documentation of physical control of evidence.

chemical agent—A toxic substance intended to be used for operations to debilitate, immobilize, or kill military or civilian personnel.

chemical ammunition—A munition, commonly a missile, bomb, rocket, or artillery shell, designed to deliver chemical agents.

chemical attack—The intentional release of toxic liquid, gas, or solid in order to poison the environment or people.

chemical warfare—The use of toxic chemicals as weapons, not including herbicide used to defoliate battlegrounds or riot control agents such as gas or mace.

chemical weapons—Weapons that produce effects on living targets via toxic chemical properties. Examples would be sarin, VX nerve gas, or mustard gas.

chemterrorism—The use of chemical agents in a terrorist operation. Well-known chemical agents include sarin and VX nerve gas.

choking agent—A compounds that injures primarily in the respiratory tract (i.e., nose, throat, and lungs). In extreme cases membranes swell up, lungs become filled with liquid, and death results from lack of oxygen.

Cipro—A Bayer antibiotic that combats inhalation anthrax.

confirmed—In the context of the threat evaluation process, a water contamination incident is confirmed if there is definitive evidence that the water has been contaminated.

counterterrorism—Measures used to prevent, preempt, or retaliate against terrorist attacks.

credible—In the context of the threat evaluation process, a water contamination threat is characterized as credible if information collected during the threat evaluation process corroborates information from the threat warning.

cutaneous—Related to or entering through the skin.

cutaneous anthrax—Anthrax that is contracted via broken skin. The infection spreads through the bloodstream, causing cyanosis, shock, sweating, and finally death.

cyanide agent—Used by Iraq in the Iran war against the Kurds in the 1980s and also by the Nazis in the gas chambers of concentration camps, a cyanide agent is a colorless liquid that is inhaled in its gaseous form, while liquid cyanide and cyanide salts are absorbed by the skin. Symptoms are headache, palpitations, dizziness, and respiratory problems followed later by vomiting, convulsions, respiratory failure, unconsciousness, and eventually death.

cyberterrorism—Attacks on computer networks or systems, generally by hackers working with or for terrorist groups. Some forms of cyberterrorism include denial of service attacks, inserting viruses, or stealing data.

dirty bomb—A makeshift nuclear device that is created from radioactive nuclear waste material. While not a nuclear blast, an explosion of a dirty bomb causes localized radioactive contamination as the nuclear waste material is carried into the atmosphere, where it is dispersed by the wind.

Ebola—Ebola hemorrhagic fever (Ebola HF) is a severe, often fatal disease in nonhuman primates such as monkeys, chimpanzees, gorillas, and also in humans. Ebola has appeared sporadically since 1976, when it was first recognized.

eBomb (or e-bomb)—Electromagnetic bomb that produces a brief pulse of energy that affects electronic circuitry. At low levels, the pulse temporarily disables electronics systems, including computers, radios, and transportation systems. High levels completely destroy circuitry, causing mass disruption of infrastructure while sparing life and property.

ecotage—The portmanteau of the "eco-" prefix and "sabotage." It is used to describe illegal acts of vandalism and violence committed in the name of environmental protection.

ecoterrorism—A neologism for terrorism that includes sabotage intended to hinder activities that are considered damaging to the environment.

Euroterrorism—Associated with left-wing terrorism of the 1960s, 1970s, and1980s involving the Red Brigade, Red Army Faction, and November 17th Group, among other groups, which targeted American interests in Europe and NATO. Other groups include Orange Volunteers, Red Hand Defenders, Continuity IRA, Loyalist Volunteer Force, Ulster Defense Association, and First of October Anti-Fascist Resistance Group.

fallout—The descent to the earth's surface of particles contaminated with radioactive material from a radioactive cloud. The term can also be applied to the contaminated particulate matter itself.

Fatah—Meaning "conquest by means of jihad"; a political organization created in the 1960s and led by Yasser Arafat. With both a military and intelligence wing, it has carried out terrorist attacks on Israel since 1965. It joined the PLO in 1968. Since 9/11, the Fatah was blamed for attempting to smuggle fifty tons of weapons into Israel.

fatwa—A legal ruling regarding Islamic law.

Fedayeen Saddam—Iraq's paramilitary organization said to be an equivalent to the Nazi's "SS." The militia is loyal to Saddam Hussein and is responsible for using brutality on civilians who are not loyal to the policies of Saddam. They do not dress in uniform.

filtrate—In ultrafiltration, the water that passes through the membrane and that contains particles smaller than the molecular weight cutoff of the membrane.

frustration-aggression hypothesis—A hypothesis that every frustration leads to some form of aggression and every aggressive act results from some prior frustration. As defined by Gurr (1968): "The necessary precondition for violent civil conflict is relative deprivation, defined as actors' perception of discrepancy between their value expectations and their environment's apparent value capabilities. This deprivation may be individual or collective."

fundamentalism—Conservative religious authoritarianism. Fundamentalism is not specific to Islam; it exists in all faiths. Characteristics include literal interpretation of scriptures and a strict adherence to traditional doctrines and practices.

Geneva Protocol 1925—The first treaty to prohibit the use of biological weapons: the 1925 Geneva Protocol for the Prohibition of the Use In War of Asphyxiating, Poisonous or Other Gases and Bacteriological Methods of Warfare.

germ warfare—The use of biological agents to cause harm to targeted people either directly, by bringing the people into contact with the agents, or indirectly, by infecting other animals and plants, which would in turn cause harm to the people.

glanders—An infectious bacterial disease known to cause inflammation in horses, donkeys, mules, goats, dogs, and cats. Human infection has not been seen since 1945, but because so few organisms are required to cause disease, it is considered a potential agent for biological warfare.

grab sample—A single sample collected at a particular time and place that represents the composition of the water, air, or soil only at that time and location.

ground zero—From 1946 until 9/11, ground zero was the point directly above, below, or at which a nuclear explosion occurred or the center or origin of rapid, intense, or violent activity or change. After 9/11, the term, when used with initial capital letters, refers to the ground at the epicenter of the World Trade Center attacks.

guerrilla warfare—The term invented to describe the tactics Spain used to resist Napoleon, though the tactic itself has been around much longer. Literally, it means "little war." Guerrilla warfare features cells and utilizes no front line. The oldest form of asymmetric warfare, guerrilla warfare is based on sabotage and ambush with the objective of destabilizing the government through lengthy and low-intensity confrontation.

Hamas—A radical Islamic organization that operates primarily in the West Bank and Gaza Strip whose goal is to establish an Islamic Palestinian state in place of Israel. On the one hand, Hamas operates overtly in its capacity as a social services deliverer, but its activists have also conducted many attacks, including suicide bombings, against Israeli civilians and military targets.

hazard—An inherent physical or chemical characteristic that has the potential for causing harm to people, the environment, or property.

hazard assessment—The process of evaluating available information about a site to identify potential hazards that might pose a risk to the site characterization team. The hazard assessment results in assigning one of four levels to risk: lower hazard, radiological hazard, high chemical hazard, or high biological hazard.

hemorrhagic fever—In general, the term *viral hemorrhagic fever* is used to describe a severe multisystem syndrome wherein the overall vascular system is damaged and the body becomes unable to regulate itself. These symptoms are often accompanied by hemorrhage; however, the bleeding itself is not usually life threatening. Some types of hemorrhagic fever viruses can cause relatively mild illnesses.

Hizbollah (Hezbollah)—Meaning "the Party of God." One of many terrorist organizations that seek the destruction of Israel and of the U.S. It has taken credit for numerous bombings against civilians and has declared that civilian targets are warranted. Hizbollah claims it sees no legitimacy for the existence of Israel and that its conflict becomes one of legitimacy that is based on religious ideals.

Homeland Security Office—An agency organized after 9/11, with former Pennsylvania governor Tom Ridge heading it up. The Office of Homeland Security is at the top of approximately forty federal agencies charged with protecting the U.S. against terrorism.

homicide bombings—A term the White House coined to replace the old "suicide bombings."
incident—A confirmed occurrence that requires response actions to prevent or minimize loss of life or damage to property and/or natural resources. A drinking water contamination incident occurs when the presence of a harmful contaminant has been confirmed.
inhalation anthrax—A form of anthrax that is contracted by inhaling anthrax spores. This results in pneumonia, sometimes meningitis, and finally death.
intifada (intifadah)—(alternatively Intifadah, from Arabic "shaking off") The two intifadas are similar in that both were originally characterized by civil disobedience by the Palestinians, which escalated into the use of terror. In 1987, following the killing of several Arabs in the Gaza Strip, the first intifada began and went on until 1993. The second intifada began in September 2000, following Ariel Sharon's visit to the Temple Mount.
Islam—Meaning "submit." The faith practiced by followers of Muhammad. Islam claims more than a billion believers worldwide.
jihad—Meaning "struggle." The definition is a subject of vast debate. There are two definitions generally accepted. The first is a struggle against oppression, whether political or religious. The second is the struggle within oneself, or a spiritual struggle.
kneecapping—A malicious wounding by firearm to damage the knee joint; a common punishment used by Northern Ireland's IRA for those accused of collaborating with the British.
Koran—The holy book of Islam, considered by Muslims to contain the revelations of God to Muhammad. Also called the Qu'ran.
Laboratory Response Network (LRN)—A network of labs developed by the CDC, APHL, and FBI for the express purpose of dealing with bioterrorism threats, including pathogens and some biotoxins.
Lassa fever—An acute, often fatal, viral disease characterized by high fever, ulcers of the mucous membranes, headaches, and disturbances of the gastrointestinal system.
LD50—The dose of a substance that kills 50 percent of those infected.
links—The means (road, rail, barge, or pipeline) by which a chemical is transported from one node to another.
mindset—According to *American Heritage Dictionary*: "1. A fixed mental attitude or disposition that predetermines a person's response to and interpretation of situations; 2. an inclination or a habit." *Merriam Webster's Collegiate Dictionary* (10th ed.) defines it as "1. A mental attitude or inclination; 2. a fixed state of mind." The term dates from 1926 but apparently is not included in dictionaries of psychology.
Molotov cocktail—A crude incendiary bomb made of a bottle filled with flammable liquid and fitted with a rag wick.

monkeypox—The Russian bioweapon program worked with this virus, which is in the same family as smallpox. In June 2003, a spate of human monkeypox cases was reported in the U.S. Midwest. This was the first time that monkeypox was seen in North America, and it was the first time that monkeypox was transferred from animal to human. There was some speculation that it was a bioattack.

mullah—A Muslim, usually holding an official post, who is trained in traditional religious doctrine and law.

Muslim (also Moslem)—Followers of the teachings of Muhammad, or Islam.

mustard gas—A blistering agent that causes severe damage to the eyes, internal organs, and respiratory system. Produced for the first time in 1822, mustard gas was not used until World War I. Victims suffered the effects of mustard gas thirty to forty years after exposure.

narcoterrorism—The view of many counterterrorist experts that there exists an alliance between drug traffickers and political terrorists.

nerve agent—The Nazis used the first nerve agents: insecticides developed into chemical weapons. Some of the better-known nerve agents include VX, sarin, soman, and tabun. These agents are used because only a small quantity is necessary to inflict substantial damage. Nerve agents can be inhaled or can absorb through intact skin.

node—A facility at which a chemical is produced, stored, or consumed.

nuclear blast—An explosion of any nuclear material that is accompanied by a pressure wave, intense light and heat, and widespread radioactive fallout, which can contaminate the air, water, and ground surface for miles around.

opportunity contaminant—A contaminant that might be readily available in a particular area, even though it may not be highly toxic or infectious or easily dispersed and stable in treated drinking water.

Osama bin Laden (also spelled "Usama")—A native of Saudi Arabia, he was born the seventeenth of twenty-four sons of Saudi Arabian builder Mohammed bin Oud bin Laden, a Yemeni immigrant. Early in his career, he helped the mujahideen fight the Soviet Union by recruiting Arabs and building facilities. He hates the U.S. and apparently this is because he views the U.S. as having desecrated holy ground in Saudi Arabia with its presence during the first Gulf War. Expelled from Saudi Arabia in 1991 and from Sudan in 1996, he operated terrorist training camps in Afghanistan. His global network al Qaeda is credited with the attacks on the U.S. on September 11, 2001, the attack on the USS *Cole* in 2000, and a number of other terrorist attacks.

pathogen—Any agent that can cause disease.

pathways—The sequences of nodes and links by which a chemical is produced, transported, and transformed from its initial source to its ultimate consumer.

plague—The pneumonic plague, which is more likely to be used in connection with terrorism, is naturally carried by rodents and fleas but can be aerosolized and sprayed from crop dusters. A 1970 World Health Organization assessment asserted that in a worst case scenario, a dissemination of fifty kilograms in an aerosol over a city of five million could result in 150,000 cases of pneumonic plague, 80,000–100,000 of which would require hospitalization and 36,000 of which would be expected to die.

political terrorism—Terrorist acts directed at governments and their agents and motivated by political goals (e.g., national liberation).

possible—In the context of the threat evaluation process, a water contamination threat is characterized as possible if the circumstances of the threat warning appear to have provided an opportunity for contamination.

potassium iodide—An FDA-approved nonprescription drug for use as a blocking agent to prevent the thyroid gland from absorbing radioactive iodine.

presumptive results—Results of chemical and/or biological field testing that need to be confirmed by further lab analysis. Typically used in reference to the analysis of pathogens.

psychopath—A mentally ill or unstable person, especially one having a psychopathic personality (*see* psychopathy), according to *Webster's*.

psychopathology—The study of psychological and behavioral dysfunction occurring in mental disorder or in social disorganization, according to *Webster's*.

psychopathy—A mental disorder, especially an extreme mental disorder marked usually by egocentric and antisocial activity, according to *Webster's*.

psychotic—Of, relating to, or affected with psychosis, which is a fundamental mental derangement (as schizophrenia) characterized by defective or lost contact with reality, according to *Webster's*.

rapid field testing—Analysis of water during site characterization uses rapid field water testing technology in an attempt to tentatively identify contaminants or unusual water quality.

red teaming—As used in this text, a group exercise to imagine all possible terrorist attack scenarios against the chemical infrastructure and their consequences.

retentate—In ultrafiltration, the retentate is the solution that contains the particles that do not pass through the membrane filter. The retentate is also called the concentrate.

ricin—A stable toxin easily made from the mash that remains after processing castor beans. At one time, it was used as an oral laxative, castor oil; ricin causes diarrhea, nausea, vomiting, abdominal cramps, internal bleeding, liver and kidney failure, and circulatory failure. There is no antidote.

salmonella—An infection caused by a gram-negative bacillus, a germ of the *Salmonella* genus. Infection with this bacteria may involve only the intestinal tract or may be spread from the intestines to the bloodstream and then to other sites in the body.

Symptoms of salmonella enteritis include diarrhea, nausea, fever, abdominal pain, and fever. Dehydration resulting from the diarrhea can cause death, and the disease could cause meningitis or septicemia. The incubation period is between eight and forty-eight hours, while the acute period of the illness can hang on for one to two weeks.

sarin—A colorless, odorless gas. With a lethal dose of .5 milligrams (a pinprick-sized droplet), it is twenty-six times more deadly than cyanide gas. Because the vapor is heavier than air, it hovers close to the ground. Sarin degrades quickly in humid weather, but sarin's life expectancy increases as the temperature gets higher, regardless of how humid it is.

sentinel laboratory—A Laboratory Response Network (LRN) lab that reports unusual results that might indicate a possible outbreak and refers specimens that may contain select biological agents in reference labs within the LRN.

site characterization—The process of collecting information from an investigation site in order to support the evaluation of a drinking water contamination threat. Site characterization activities include the site investigation, field safety screening, rapid field testing of the water, and sample collection.

sleeper cell—A small cell that keeps itself undetected until such time as it can "awaken" and cause havoc.

smallpox—The first biological weapon, used during the eighteenth century, smallpox killed three hundred million people in the nineteenth century. There is no specific treatment for smallpox disease, and the only prevention is vaccination. This currently poses a problem, since the vaccine was discontinued in 1970 and the WHO declared smallpox eradicated. Incubation is seven to seventeen days, during which the carrier is not contagious. Thirty percent of people exposed are infected, and it has a 30 percent mortality rate.

sociopath—Basically synonymous with psychopath (q.v.). Symptoms in the adult sociopath include an inability to tolerate delay or frustration, a lack of guilt feelings, a relative lack of anxiety, a lack of compassion for others, a hypersensitivity to personal ills, and a lack of responsibility. Many authors prefer the term *sociopath* because this type of person had defective socialization and a deficient childhood.

sociopathic—Of, relating to, or characterized by asocial or antisocial behavior or a psychopathic (q.v.) personality, according to *Webster's*.

spore—An asexual, usually single-celled reproductive body of plants such as fungi, mosses, or ferns; a microorganism, as a bacterium, in a resting or dormant state.

Strategic National Stockpile—A stock of vaccines and antidotes stored at the Centers for Disease control in Atlanta, to be used against biological warfare.

terrorist group—A group that practices or has significant elements that are involved in terrorism.

threat—An indication that a harmful incident, such as contamination of the drinking water supply, may have occurred. The threat may be direct, such as a verbal or written threat, or circumstantial, such as a security breach or unusual water quality.
toxin—A poisonous substance produced by living organisms capable of causing disease when introduced into the body tissues.
transponder—A device on an airliner that sends out a signal allowing air traffic controllers to track the aircraft. Transponders were disabled in some of the planes hijacked 9/11.
Transportation Security Administration (TSA)—An agency created by the Patriot Act of 2001 for the purpose of overseeing technology and security in American airports.
tularemia—An infectious disease caused by a hardy bacterium, *Francisella tularensis*, found in animals, particularly rabbits, hares, and rodents. Symptoms depend upon how the person was exposed to tularemia but can include difficulty breathing, chest pain, bloody sputum, swollen and painful lymph glands, ulcers on the mouth or skin, swollen and painful eyes, and sore throat. Symptoms usually appear from three to five days after exposure but sometimes take up to two weeks. Tularemia is not spread from person to person, so people who have it need not be isolated.
ultrafiltration—A filtration process for water that uses membranes to preferentially separate very small particles that are larger than the membrane's molecular weight cutoff, typically greater than 10,000 daltons. (A dalton is a unit of mass, defined as one-twelfth the mass of a carbon-12 nucleus. It is also called the atomic mass unit, abbreviated as either "amu" or "u").
vector—An organism that carries germs from one host to another.
vesicle—A blister filled with fluid.
weapons of mass destruction (WMD)—According to the National Defense Authorization Act, any weapon or device that is intended, or has the capability, to cause death or serious bodily injury to a significant number of people through the release, dissemination, or impact of (a) toxic or poisonous chemicals or their precursors, (b) a disease organism, or (c) radiation or radioactivity.
xenophobia—Irrational fear of strangers or those who are different from oneself.
zyklon b—A form of hydrogen cyanide. Symptoms of inhalation include increased respiratory rate, restlessness, headache, and giddiness followed later by convulsions, vomiting, respiratory failure, and unconsciousness. Used in the Nazi gas chambers in World War II.

REFERENCES AND RECOMMENDED READING

Bergen, P. L. 2002. *Holy War, Inc: Inside the secret world of Osama bin Ladin.* New York: Touchstone Press.

CRS. 2006. *Chemical Facility Security. CRS Report for Congress.* Washington, DC: Congressional Research Service—The Library of Congress.

Excerpts from statements in court. 1998. *New York Times.* January 9, B4.

FBI. 2006. *Terrorism 2002–2005.* Washington, DC: Federal Bureau of Investigation. www.fbi.gov/publications/terror/terrorism2000_2005.htm (accessed January 5, 2010).

Gurr, T. R. 1968. Psychological factors in civil violence. *World Politics* 20: 245–78.

Haimes, Y. Y. 2004. *Risk modeling, assessment, and management.* 2nd ed. New York: John Wiley & Sons, 699.

Henry, K. 2002. The face of homeland security. *Government Security,* Apr. 1, 30–37.

Lindsey, H. 2001. *Vocabulary of hate.* www.worldnetdaily.com (accessed April 18, 2008).

Meyer, E. 2005. *Chemistry of Hazardous Materials.* 4th ed. NY: Prentice Hall.

Napolitano tells it like it isn't. 2009. *Washington Times.* www.washingtontimes.com/news/2009/mar/29/tell-it-like-it-is-man-caused-disasters-is-napolit/ (accessed March 29, 2009).

NSHS. 2006. *National Strategy for Homeland Security.* www.whitehouse.gov/homeland (accessed May 13, 2006).

Old Dominion University. 2000; 2002. *Violence in the workplace: Security concerns.* From a series of lectures presented to environmental health students. Norfolk, VA.

OSHA. 2007. *Combustible dust national emphasis program: CPL 03-00-006.* Washington, DC: U.S. Department of Labor. www.osha.gov/pls/oshaweb/owadisp.show_document?p_table=DIRECTIVES&p_id=3729 (accessed April 14, 2008).

Perl, R. 2004. *Terrorism and national security: Issues and trends.* CRS Issue Brief IB10119. Washington, DC.

Sauter, M. A., and J. J. Carafano. 2005. *Homeland security: A complete guide to understanding, preventing, and surviving terrorism.* New York: McGraw-Hill.

Spellman, F. R. 1997. *A guide to compliance for process safety management/risk management planning (PSM/RMP).* Lancaster, PA: Technomic.

U.S. Congress. 2005. *Annual country reports on terrorism.* 22 USC, Chapter 38, Section 2656f.

2

Critical Infrastructure

Grand Coulee Dam, Columbia River, Washington

While it is not so easy to definitively define terrorism and/or the terrorist, we have less difficulty identifying the likely targets of terrorists. In America, we call these likely targets our critical infrastructure.

"So in war, the way is to avoid what is strong and to strike at what is weak."

—*Sun Tzu*

WHAT IS CRITICAL INFRASTRUCTURE?

For the United States of America, 9/11 was a slap in the face, a punch in the gut (actually, the ultimate sucker punch), and a most serious wake-up call. The 9/11 wake-up call generated several reactions on our part—obviously, protecting ourselves from further attack became (and hopefully still is) priority number one. In light of this important need (i.e., the survival of our way of life), the Department of Homeland Security was created. According to Governor Tom Ridge, "You may say Homeland Security is a Y2K problem that doesn't end Jan. 1 of any given year" (Henry 2002). And, according to Barack Obama (2007), the Department of Homeland Security does "the work that ensures no other family members have to lose a loved one to a terrorist who turns a plane into a missile, a terrorist who straps a bomb around her waist and climbs aboard a bus, a terrorist who figures out how to set off a dirty bomb in one of our cities."

Among other things, the new emphasis on homeland security pointed to the need to protect and enhance the security of the nation's critical infrastructure. Critical infrastructure can be defined or listed in many ways. Generally, governments use the term to describe material assets, systems, and services that are essential for the functioning of an economy and society and maintaining public confidence. Destruction or compromise of any of these systems or services would have a debilitating impact on the area either directly, through interdependencies, or from cascading effects. For the purpose of this text, critical infrastructure is defined as those assets of physical, key resources and computer-/service-based systems that are essential to the minimum operations of economy and government. Critical infrastructures (in the authors' opinion) are the following:

- agriculture
- banking and finance
- chemical and hazardous materials
- commercial assets
- dams
- defense industrial base
- emergency services
- energy
- information technology
- national monuments and icons
- nuclear power plants
- organizations
- postal and shipping
- public health
- strategies and assessments
- telecommunications
- transportation
- water and water treatment systems

Although we did not list cyberspace and all ancillaries (excluding, of course, the listing of information technology) involved in or with digital operations (e-technology), in this current era we can state without equivocation, doubt, ambiguity, or vagary that the digital connection is the glue that holds all critical infrastructure together. This is

DID YOU KNOW?

Over 85 percent of the critical infrastructure within the U.S. is owned and operated by the private sector.

the case, of course, because all separate infrastructures are interconnected in one way or another. This may surprise you to some degree, but think about it—we are not speaking about rocket science here; instead we are speaking about the present reality of e-technology. It would be hard to imagine that any of the above listed infrastructure sectors could operate today without e-technology.

For example, consider e-agriculture. The Food and Agriculture Organization of the United Nations (FAO 2005) defines e-agriculture as "an emerging field in the intersection of agricultural informatics, agricultural development entrepreneurship, referring to agriculture services, technology dissemination, and information delivered or enhanced through the Internet and related technologies. More specifically, it involves the conceptualization, design, development, evaluation and application of new (innovative) ways to use existing or emerging information and communication technologies."

DID YOU KNOW?

More advanced applications of e-agriculture in farming exist in the use of sophisticated information and communication technologies such as satellite systems, global positioning systems (GPS), and advanced computers and electronic systems to improve the quantity and quality of production (FAO 2005).

Even though we did it in the past, today how would we go about withdrawing money from the bank without e-banking technology? Today we can conduct our banking at any time we wish, from any location in the world. Modern life without debit cards and ATMs would be a rude awakening for many of us.

We could go down the list of critical infrastructures and easily point out where e-technology and the specific industry interface. However, since this is a discussion focusing on the energy industry, we will keep the focus on e-technology as it relates to or interfaces with the energy industry. If not familiar with the energy industry—an oil refinery, hydroelectric dam, or wind farm, for example—it might surprise you to know that the entire modern oil refining process and electrical distribution grid is operated by e-technology from computer operation stations (manned by people, of course) with various digital proximity switches and other devices strategically positioned throughout the process to operate valves and switches, monitor critical parameters, and provide automatic emergency shutdown procedures.

ENERGY SECTOR INFRASTRUCTURE*

In the first three volumes of our critical infrastructure series, *Water Infrastructure Protection and Homeland Security, Food Infrastructure Protection and Homeland Security,* and *The Chemical Industry and Homeland Security,* the message/focus was on (as the titles suggest) water/wastewater, agriculture, and chemical manufacturing, processing, and storage. In this fourth volume of the series, *Energy Infrastructure Protection and Homeland Security,* even though the target is different—pointing out and discussing the threat to our energy industry—we use the same proven format. In addition, this text describes the study, design, and implementation of precautionary measures aimed to reduce the risk to our energy infrastructure from both homegrown and/or foreign terrorism.

U.S. Department of Homeland Security (2009) points out that energy infrastructure fuels the economy of the twenty-first century. Without a stable energy supply, health and welfare are threatened and the U.S. economy cannot function.

The energy infrastructure is divided into three interrelated segments: electricity, petroleum, and natural gas. Each of these segments is discussed in detail in chapter 3, but for now a brief overview of each segment is provided.

Electricity Segment

The U.S. electricity segment contains more than 5,300 power plants with approximately 1,075 gigawatts of installed generating capacity. Approximately 49 percent of electricity is produced by combusting coal (primarily transported by rail), 19 percent in nuclear power plants, and 20 percent by combusting natural gas. The remaining generation is provided by hydroelectrical plants (7 percent), oil (2 percent), and renewable (solar, wind, and geothermal) and other sources (3 percent). Electricity generated at power plants is transmitted over 211,000 miles of high-voltage transmission lines. Voltage is stepped down at substations before being distributed to 140 million

*This section is taken from the Department of Homeland Security's *National Infrastructure Protection Plan: Energy Sector*, www.dhs.gov/nipp (accessed 2009).

customers over millions of miles of lower-voltage distribution lines. The electricity infrastructure is highly automated and controlled by regional grid operators using sophisticated energy management systems that are supplied by supervisory control and data acquisition (SCADA) systems to keep the system in balance.

Petroleum Segment

The petroleum segment entails the exploration, production, storage, transport, and refinement of crude oil. The crude oil is refined into petroleum products that are then stored and distributed to key economic sectors throughout the United States. Key petroleum products include motor gasoline, jet fuel, distillate fuel oil, residual fuel oil, and liquefied petroleum gases. Both crude oil and petroleum products are imported, primarily by ship, as well as produced domestically. Currently, 66 percent of the crude oil required to fuel the U.S. economy is imported. In the United States, there are more than 500,000 crude oil–producing wells, 30,000 miles of gathering pipeline, and 51,000 miles of crude oil pipeline. There are 133 operable petroleum refineries, 116,000 miles of product pipeline, and 1,400 petroleum terminals. Petroleum also relies on sophisticated SCADA and other systems to control production and distribution; however, crude oil and petroleum products are stored in tank farms and other facilities.

Natural Gas Segment

Natural gas is also produced, piped, stored, and distributed in the Unites States. Imports of liquefied natural gas (LNG) are increasing to meet growing demand. There are more than 448,000 gas production and condensate wells and 20,000 miles of gathering pipeline in the country. Gas is processed (impurities removed) at more than 550 gas processing plants, and there are almost 302,000 miles of interstate and intrastate pipeline for the transmission of natural gas. Gas is stored at 399 underground storage fields and 103 LNG peak shaving facilities. Finally, natural gas is distributed to homes and businesses over 1,175,000 miles of distribution pipelines. The heavy reliance on pipelines highlights the interdependency with the transportation sector, and the reliance on the energy sector for power means that virtually all sectors have dependencies with the energy sector.

THE BOTTOM LINE

Again, it is important to point out that the energy sector provides products and materials that are essential to the U.S. economy and to the so-called good life (standard of living) we presently enjoy. The energy sector is well aware of its vulnerabilities and is leading a significant voluntary effort to increase its planning and preparedness. Cooperation through industry groups has resulted in substantial information sharing of

HSPD-7 (PROTECTING CRITICAL INFRASTRUCTURE)

In terms of protecting critical infrastructure, agriculture was added to the list in December 2003 by Homeland Security Presidential Directive 7 (HSPD-7), "Critical Infrastructure Identification, Prioritization, and Protection." This directive instructs agencies to develop plans to prepare for and counter the terrorist threat. HSPD-7 mentions the following industries: agriculture and food; banking and finance; transportation (air, sea, and land, including mass transit, rail, and pipelines); energy (electricity, oil, and gas); telecommunications; public health; emergency services; drinking water; and water treatment (wastewater treatment is implied).

effective and best practices across the sector. Many sector owners and operators have extensive experience with infrastructure protection and have more recently focused their attention on cyber security. In addition to the economic consequences of a successful homegrown or foreign terrorist attack against energy sector facilities, there is also the potential of a threat to public health and safety and to the environment. We hope that this book and the others in the critical infrastructure series will aid in the prevention and mitigation of deliberate attacks.

REFERENCES AND RECOMMENDED READING

Breeze, R. 2004. Agroterrorism: Betting far more than the farm. *Biosecurity and Bioterrorism: Biodefense Strategy, Practice and Science* 2(4): 1–14.

Carus, S. 2002. *Bioterrorism and biocrimes: The illicit use of biological agents since 1900.* Washington, DC: Center for Counterproliferation Research, National Defense University.

CBO. 2004. *Homeland security and the private sector.* Washington, DC: Congressional Budget Office.

Chalk, P. 2004. *Hitting America's soft underbelly: The threat of deliberate biological attacks against the U.S. agriculture and food industry.* Santa Monica, CA: RAND.

FAO. 2005. *Bridging the rural digital divide.* New York: United Nations Food and Agriculture Organization. www.fao.org/rdd (accessed April 19, 2008).

FR. 2003. Notice of proposed rulemaking. *Federal Register* 68(90).

Henry, K. 2002. The face of homeland security. *Government Security,* Apr. 1, 30–37.

Horn, F. P. 1999. Statement made before the United States Senate Emerging Threats and Capabilities Subcommittee of the Armed Services Committee. www.Senate.gov/~armed _services/statemnt/1999/991027fh.pdf (accessed June 27, 2007).

Lane, J. 2002. Sworn testimony. Congressional Field Hearing, House Committee on Government Reform, Abilene, KS.

Obama, B. 2007. *Homeland security.* http://www.whitehouse.gov/agenda/homeland_security/.

Parker, H. S. 2002. *Agricultural bioterrorism: A federal strategy to meet the threat.* McNair Paper 65, National Defense University. www.ndu.edu/inss/docuploaded/McNair65.pdf.

U.S. Department of Homeland Security. 2009. *National infrastructure protection plan: Energy sector.* www.dhs.gov/nipp (accessed April 24, 2009).

3

The Energy Sector

Chief Joseph Dam, Columbia River, Washington

INTRODUCTION*

The United States energy sector is vital to the U.S. economy. It is a high-tech, research and development (R&D)–oriented industry that, as mentioned, includes assets related to three key energy resources: electrical power, petroleum, and natural gas. Each of these resources requires a unique set of supporting activities and assets, as shown in table 3.1. Petroleum and natural gas share similarities in methods of extraction, fuel cycles, and transport, but the facilities and commodities are separately regulated and have multiple stakeholders and trade associations.

Energy assets and critical infrastructure components are owned by private, federal, state, and local entities, as well as by some types of energy consumers, such as large industries and financial institutions (often for backup-power purposes). Types of major asset ownership are shown in table 3.2.

---

*Much of the information in this section is taken from *Energy: Critical Infrastructure and Key Resources Sector-Specific Plan as Input to the National Infrastructure Protection Plan* (Washington, DC Department of Homeland Security, 2007).

DID YOU KNOW?

The Department of Energy is responsible for the management of the extensive National Laboratory System, which represents one of the most comprehensive research enterprises in the world. These laboratories perform research and development that is multidisciplinary in nature and for which there is a strong public and national purpose (USDOE 2009).

ELECTRICITY

The electricity portion of the energy sector includes the generation, transmission, and distribution of electricity (see figure 3.1). The Energy Information Administration (2009) points out that the use of electricity is ubiquitous, spanning all sectors of the U.S. economy. Electricity generation accounted for 40 percent of all energy consumed in the United States in 2009. Although there are some significant regional differences, more than 98 percent of electricity is generated domestically, though some of the fuels used to generated electricity are imported.

Electricity systems facilities are dispersed throughout the North American continent. Although most assets are privately owned, no single organization represents the interests of the entire sector. The North American Electric Reliability Corporation (NERC), through its eight regional reliability councils, provides a platform for ensuring reliable, adequate, and secure supplies of electricity through coordination with many asset owners.

Table 3.1. Segments of the Energy Sector

Electricity	*Petroleum*	*Natural Gas*
▪ Generation	▪ Crude oil	▪ Production
▪ Transmission	▪ Petroleum processing facilities	▪ Transport
▪ Distribution		▪ Distribution
▪ Control systems		▪ Storage
▪ Electricity markets		▪ Liquefied natural gas
		▪ Control systems
		▪ Gas markets

Note: Hydroelectric dams, nuclear facilities, and rail and pipeline transportation are covered in other sectors.

Table 3.2. Major Asset Ownership

Ownership Entities	*Assets*
Federal government	Hydroelectric dams, nuclear and fossil-fuel power generation stations, and high-voltage transmission
State and local government	All municipal utilities
Regulated utilities	Most of the electric and natural gas infrastructure in the U.S., including major interstate pipeline companies, hydroelectric facilities, storage facility operators, and LNG terminal owners
Unregulated energy companies	Energy infrastructure assets, such as merchant generation companies owning power plants that participate in wholesale power markets
Unregulated non-energy companies	Generation plants, refineries, and oil and gas production facilities
Cooperatives	Generation, transmission, and distribution
Foreign entities	Several utilities and power stations

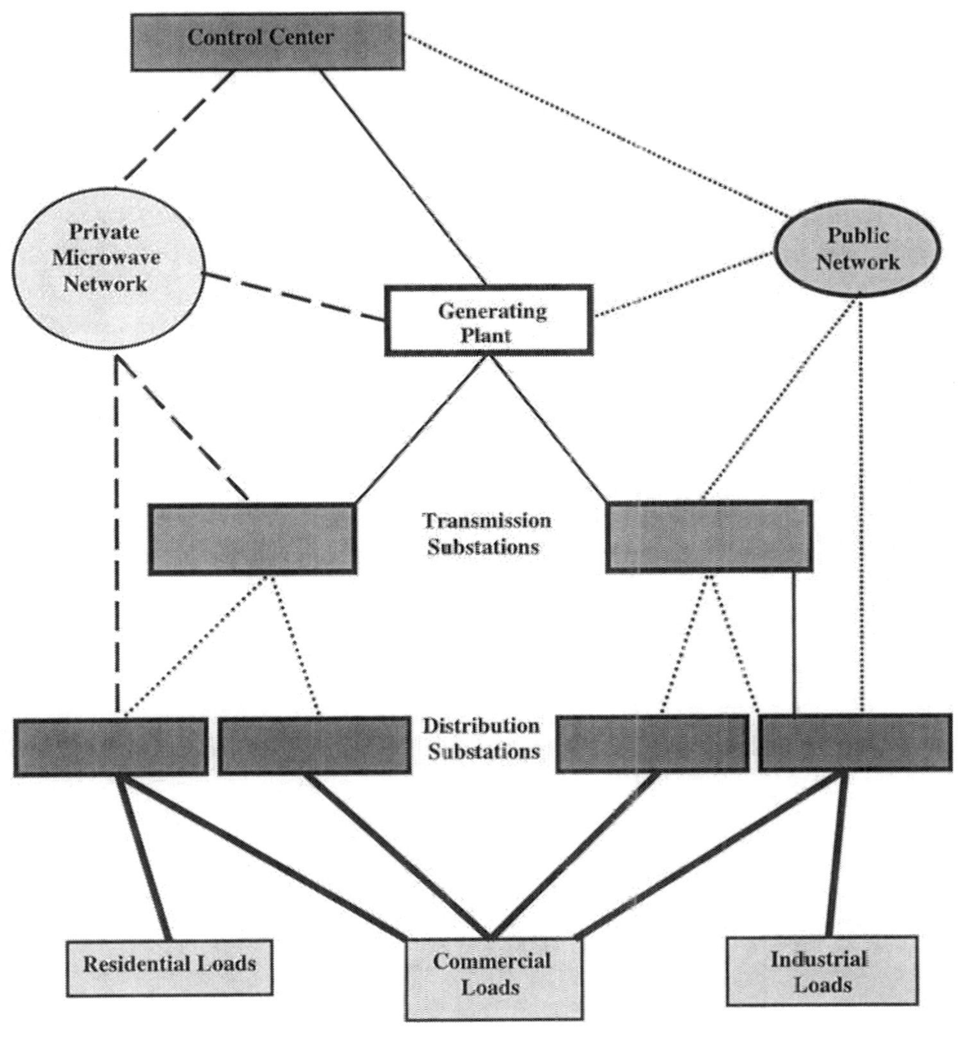

FIGURE 3.1
Schematic of Electric Power System and Control Communications

DID YOU KNOW?

NERC was founded as a nonprofit organization in 1968. It was designated as the Electric Reliability Organization (ERO) by the Federal Energy Regulatory Commission following passage of the Energy Policy Act of 2005. As a result of the law, NERC's official name changed to the North American Electric Reliability Corporation. The ERO will develop and enforce mandatory reliability standards for the bulk electric power systems in the United States, Canada, and a portion of Baja, Mexico.

Electricity Generation

The burning of fossil fuels (coal, natural gas, and oil) provides more than 70 percent of the electricity generated in the United States, as shown in table 3.3. Virtually all coal is mined domestically and then transported to power plants by rail and barge. Natural gas and oil are transported to power plants by pipeline.

Non-hydropower renewable energy sources (e.g., wind, solar, geothermal) account for a small but growing percentage of national electricity generations, with the potential to provide alternative power sources for critical facilities and functions.

Transmission Lines

Transmission lines serve two primary purposes: They move electricity from generation sites to customers, and they interconnect systems. Voltages in the transmission

Table 3.3. Existing Capacity by Energy Source (Megawatts) (2007)

Energy Source	*Number of Generators*	*Generator Capacity*
Coal	1,470	336,040
Petroleum	3,743	62,394
Natural gas	5,439	449,389
Other gases	105	2,663
Nuclear	104	105,764
Hydroelectric	3,992	77,644
Wind	389	16,596
Solar thermal and photovoltaic	38	503
Wood and wood-derived fuels	346	7,510
Geothermal	224	3,233
Other biomass	1,299	4,834
Pumped storage	151	20,355
Other	42	866
Total	17,342	1,087,791

Source: Energy Information Administration, "Existing Capacity by Energy Source," EIA, www.eia.doe.gov/cneaf/electricity/epa/epat2p2.html (accessed April 26, 2009).

system are high, which makes it possible to carry electrical power efficiently over long distances and deliver it to substations near customers.

Transmission and Distribution Substations

Substations are located at the ends of transmission lines. A transmission substation located near a power plant uses large transformers to increase the voltage to higher levels. At the other end of the transmission line, a substation uses transformers to step transmission voltages back down to distribution voltages so that the electricity can be distributed to customers.

Control Centers

Control centers have sophisticated monitoring and control systems and are staffed by operators twenty-four hours per day, 365 days per year. These operators are responsible for several key functions, including balancing power generation and demand, monitoring flows over transmission lines to avoid overloading, planning and configuring the system to operate reliably, maintaining system stability, preparing for emergencies, and placing equipment out of service and back into service for maintenance and emergencies.

Distribution Lines

Distribution lines (see figure 3.2) carry electricity from substations to end users.

FIGURE 3.2
Electrical Distribution Towers and Lines

Control Systems

Supervisory control and data acquisition systems (SCADA) and other control systems monitor the flow of electricity from generators through transmission and distribution lines. These electronic systems enable efficient operation and management of electrical systems through the use of automated data collection and equipment control.

> **IMPORTANT POINT!**
>
> The safeguarding of supervisory control and data acquisition (SCADA) systems is a paramount cause for concern for the national security of our country and for its security professionals.

PETROLEUM

The petroleum portion of the energy sector includes the production, transportation, and storage of crude oil; the processing of crude oil into petroleum products; the transmission, distribution, and storage of petroleum products; and sophisticated control systems to coordinate storage and transportation (see figure 3.3 and table 3.4).

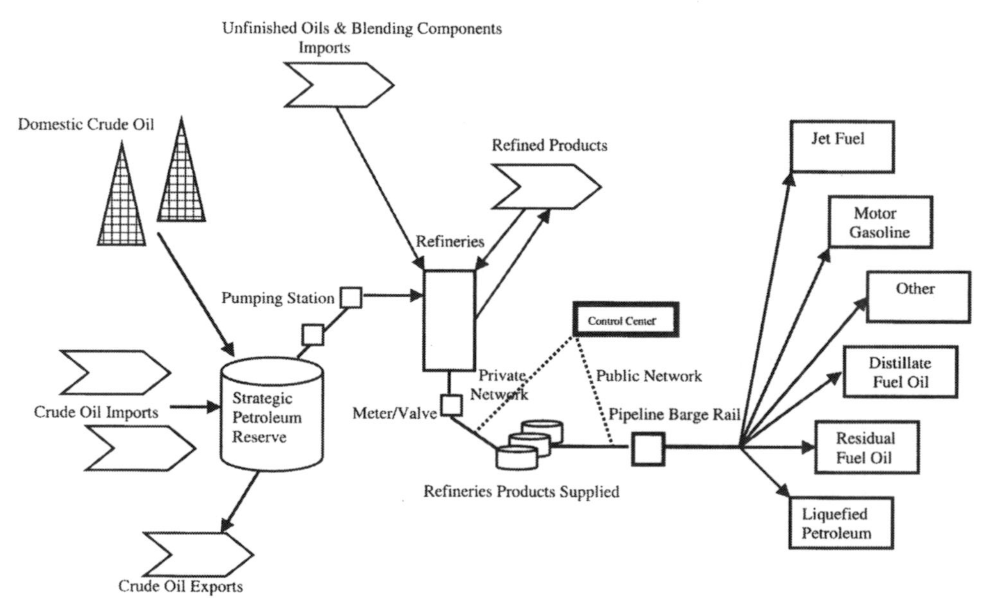

FIGURE 3.3
Schematic of the Petroleum System

Table 3.4. Petroleum Statistics for 2005 (EIA 2009)

Function	*Data*
Production	506,000 producing wells
Gathering	> 30,000 miles of gathering pipeline
Processing	149 petroleum refineries
Storage	1,400 petroleum terminals
Transportation	66% pipelines; 27% water carriers; 4% motor carriers; 2% railroads
Pipelines	284 billion ton miles of crude pipelines; 316 billion ton miles of product pipelines

Petroleum accounted for 40 percent of U.S. energy consumption in 2005. Its primary use is in the transportation systems sector, where it accounts for 98 percent of energy consumption. Petroleum is used to lesser degrees in other sectors, accounting for 30 percent of energy used in the industrial sector, 7 percent in the residential, 4 percent in the commercial, and 3 percent in the electrical power sectors (EIA 2009). It is important to note that this text does not address the chemical industry and the overlap between the petrochemical industry and the transportation, storage, and processing of crude oil and refined petroleum products.

DID YOU KNOW?

Pipelines, which are critical for the gathering, transmission, and distribution of petroleum and natural gas, are part of the transportation sector, and oversight of pipeline security is the responsibility of DHS's Transportation Security Administration (TSA).

Crude Oil (EIA 2009)

- Onshore and offshore fields—U.S. crude oil production is concentrated onshore and offshore along the Texas-Louisiana Gulf Coast, extending inland through west Texas, Oklahoma, and eastern Kansas. There are also significant oil fields in Alaska along the central North Slope. U.S. proved crude oil reserves (i.e., reserves believed to be recoverable from known reservoirs under existing economic and operating conditions) totaled an estimated 21.8 billion barrels at the close of 2005. More than three-quarters (80 percent) of U.S. reserves are in Alaska, California, Texas, and offshore areas. Petroleum production from the Alaskan North Slope is now equaled by output from the offshore areas in the federal domain seaward of the coastline along California and the western and central coasts of the Gulf of Mexico.

- Crude oil drilling, gathering, and processing—The upstream sector of the petroleum industry includes a large number of facilities, such as wellheads, gas and oil separation plants, oil/gas dehydration units, emulsion breaker units, oil/gas sweetening units, compressor stations, water treatment units, and so on for both onshore and offshore areas.
- Import marine terminals—The United States' dependence on foreign crude oil has grown from 15 percent in 1971 to 66 percent in 2005. Crude oil is received into the U.S. at import terminals, which usually consist of a berth or port facility for the tankers, unloading facilities, storage facilities, and a system of pipelines to move the crude.
- Crude oil transport—Privately owned pipelines transport most of the crude oil in the United States. Waterborne transportation modes, including ocean tankers and barges, are also used.
- Crude oil storage—Import terminals always incorporate storage facilities. At the end of 2005, U.S. crude oil inventories, including the Strategic Petroleum Reserve (SPR), totaled 1,008 million barrels. More than two thirds is stored in huge underground salt caverns at the SPR along the coastline of the Gulf of Mexico. The reserve has the capacity to hold 727 million barrels and is the world's largest supply of emergency crude oil.

Petroleum Processing, Product Transport, Storage, and Control

- Refineries—Refineries process crude oil into petroleum products such as gasoline, diesel fuel, jet fuel, and home heating oil. The Gulf Coast has more than twice the crude oil distillation capacity of any other U.S. region.
- Petroleum product transport—Petroleum products are mainly transported by pipeline, tanker, or barge, but railroad tank cars or trucks are also used. The products are shipped to terminals for temporary storage before transport to smaller bulk plants in market areas.
- Petroleum product storage—Petroleum products are stored both above and below ground in tank farms and storage fields to minimize unwanted fluctuations in pipeline throughput and product delivery. DOE's Northeast Home Heating Oil Reserve stores two million barrels of home heating oil at commercial terminals in the Northeast. This oil is intended for distribution during severe heating oil supply disruptions in that part of the country.
- Petroleum control systems—Control systems continuously monitor, transmit, and process pipeline data (e.g., flow rate, pressure, speed). SCADA systems monitor and control pumping stations and track terminal inventories.

NATURAL GAS

The natural gas portion of the energy sector includes the production, processing, transportation, distribution, and storage of natural gas, liquefied natural gas (LNG) facilities, and gas control systems (see figure 3.4).

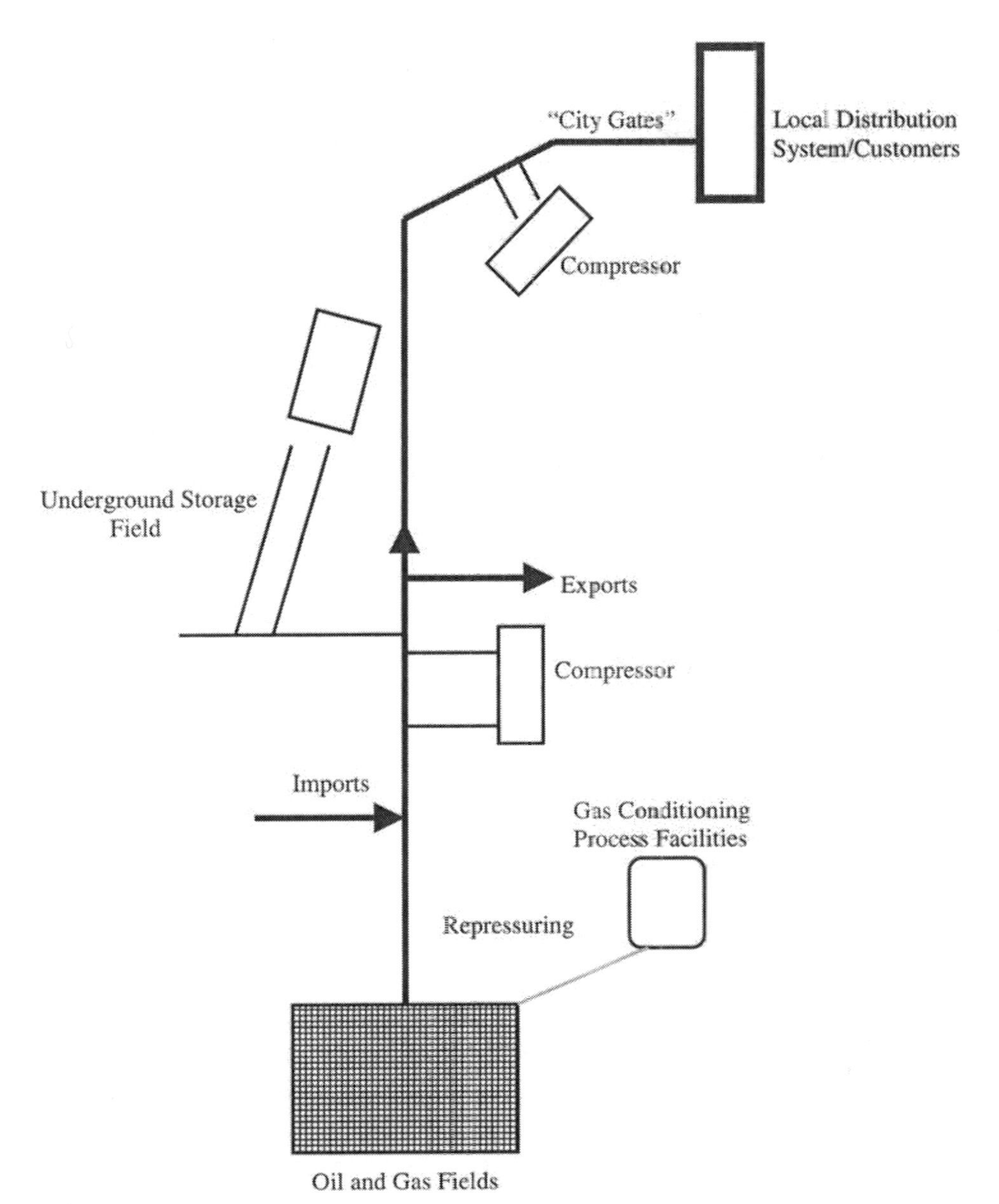

FIGURE 3.4
Schematic Showing Flow of Natural Gas

The U.S. produces roughly 20 percent of the world's natural gas supply. There are 278,000 miles of natural gas pipelines and 1,119,000 miles of natural gas distribution lines in the U.S. (Department of Homeland Security 2003). Natural gas provided 23 percent of U.S. energy needs in 2005, and its use is growing. In particular, power producers and industrial facilities are opting for gas-powered equipment, and residential customers use natural gas for heating and cooking (EIA 2009).

Although most of the gas consumed in the U.S. is produced domestically, imports have increased. This trend is likely to continue over the next few years as imported LNG assumes a larger role in the market supply.

Natural Gas Production, Processing, Transport, Distribution, and Storage

- Natural gas production—Federal Offshore Gulf of Mexico and Texas are the largest gas-producing regions in the U.S., at approximately eleven billion and thirteen billion cubic feet per day, respectively. The two regions account for almost half of all U.S. natural gas production. The United States had 193 trillion cubic feet of dry natural gas reserves as of December 31, 2004 (EIA 2009).
- Natural gas processing—Natural gas processing consists of separating all of the various hydrocarbons and fluids from the pure natural gas to produce pipeline-quality dry natural gas. Most U.S. natural gas processing plants are located near production facilities in the Southwest and Rocky Mountain states. The natural gas extracted from a well is transported to a processing plant through a network of gathering pipelines.
- Natural gas transportation—The interstate natural gas pipeline network transports natural gas from processing plants in producing regions to areas with high natural gas requirements, particularly large urban areas. Compression stations along the pipeline transmission route keep the gas moving at the desired pressure.
- Natural gas distribution—Local distribution companies typically transport natural gas from interstate pipeline delivery points to end users through thousands of miles of distribution pipe. Delivery points for local distribution companies are often termed *city gates*, especially for large municipal areas, and are important marker centers for the pricing of natural gas.
- Natural gas storage—Gas is typically stored underground and under pressure as an efficient way to balance discrepancies between supply input and market demand. Three types of facilities are used for underground gas storage: depleted reservoirs in oil and/or gas fields, aquifers, and salt caverns. Facilities serving the interstate market are subject to Federal Energy Regulatory Commission (FERC) regulations; otherwise, they are state regulated. Most working gas held in storage facilities is held under lease with shippers, local distribution companies, or end users who own the gas.

Liquefied Natural Gas Facilities

LNG is produced by cooling natural gas to –260 degrees Fahrenheit (–160 degrees Celsius). In its liquid state, natural gas occupies 618 times less volume than the same mass of gaseous methane at standard conditions, which allows it to be transported by specially designed ships or tankers. The lower forty-eight states have five marine terminals for receiving, storing, and regasifying LNG for delivery into the pipeline network and more than fifty above-ground LNG storage tanks for meeting peak-day demand.

Natural Gas Control Systems

To monitor and control the flow of natural gas, centralized control stations collect, assimilate, and manage data received from compressor stations all along the pipeline. These control systems can integrate gas flow and measurement data with other accounting, billing, and contract systems.

Gas Market Centers

Currently, thirty-seven natural gas market centers operate in the United States and Canada. These centers provide gas shippers with many of the physical capabilities and administrative support services formerly handled by interstate pipeline companies as bundled sales services (e.g., physical coverage of short-term receipt/delivery balancing needs). These centers have developed new and unique Internet-based access to gas trading platforms and capacity release programs; provide title transfer services between parties that buy, sell, or move their gas through the centers; and offer connections with other pipelines and access to storage services. These markets and their information systems are critical components of the natural gas infrastructure.

ENERGY SECTOR: A REGULATED ENVIRONMENT

Even before 9/11, the energy sector was (and is) one of the nation's most regulated industries. It is subject to numerous environmental regulations as well as the voluntary obligations imposed by the energy sector's environmental, health, and safety improvement initiatives. Including federal/state OSHA statutes, major federal statutes, as well as numerous state laws, impose significant compliance and reporting requirements on the sector (see sidebar 3.1).

Sidebar 3.1. Major Health, Safety, and Environmental Legislation for Energy Sector

1. **Energy Policy and Conservation Act of 1975** began the modern energy policy era, which, not coincidentally, was a response to an oil price spike after an OPEC embargo in 1973–1974. The act not only created the Strategic Petroleum Reserve to counter severe disruptions in the nation's oil supply, but also introduced for the first time Corporate Average Fuel Economy (CAFE) standards for automobile manufacturers, requiring that average fuel economy of vehicles sold by auto manufacturers in the U.S. achieve double in fuel efficiency.
2. **Clean Air Act (CAA)** was first passed in 1955 as the Air Pollution Control, Research and Technical Assistance Act and amended in 1963 to become the CAA. A more significant statute was passed in 1970 and amended in 1977 and 1990. It provides the EPA authority to regulate air pollutants from a wide variety

of sources, including automobiles, electrical power plants, chemical plants, and other industrial sources.

3. **Clean Water Act (CWA)** was first enacted in 1948 as the Federal Water Pollution Control Act. Subsequent extensive amendments defined the statute to be known as the CWA in 1972; it was further amended in 1977 and 1987. The CWA provides the EPA authority to regulate effluents from sewage treatment works, refineries, chemical plants, and other industry sources into U.S. waterways. The EPA has recently undertaken control efforts in on-point source pollution as well.
4. **Comprehensive Environmental Response Compensation and Liability Act of 1980 (CERCLA) and the Superfund Amendments and Reauthorization Act of 1986 (SARA)** provide the basic legal framework for the federal "Superfund" program to clean up abandoned hazardous waste sites.
5. **Emergency Planning and Community Right-to-Know Act of 1986**, also known as SARA Title III, mandates state and community development of emergency preparedness plans and also establishes an annual manufacturing-sector emissions reporting program.
6. **Resource Conservation and Recovery Act (RCRA) of 1976** provides the EPA with authority to establish standards and regulations for handling and disposing of solid and hazardous wastes (cradle-to-grave provisions).
7. **Occupational Safety and Health Act (OSH Act) of 1970** provides the Department of Labor authority to set comprehensive workplace safety and health standards, including permissible exposures to chemicals in the workplace, and authority to conduct inspections and issue citations for violations of safety and health regulations.
8. **Safe Drinking Water Act**, enacted in 1974 and amended in 1977 and again in 1986, establishes standards for public drinking water supplies.
9. **Hazardous Materials Transportation Act (HMTA)** provides the Department of Transportation the authority to regulate the packaging and movement of hazardous materials.
10. **Pollution Prevention of Act of 1990** makes it the national policy of the United States to reduce or eliminate the generation of waste at the source whenever feasible and directs the EPA to undertake a multimedia program of information collection, technology transfer, and financial assistance to the states to implement this policy and to promote the use of source reduction techniques.
11. **Energy Policy Act of 1992** introduced one of the key drivers of the renewable energy industry to date, the Production Tax Credit (PTC), which offers to independent power producers a subsidy per kilowatt hour generated from renewable sources for a period of ten years from the beginning of power generation.
12. **Federal Energy Regulatory Commission (FERC)** is an independent regulatory agency within the Department of Energy and is perhaps the leading regulatory body in determining the impact on the average consumer of energy. FERC has jurisdiction over electricity pricing, licensing for hydroelectric plants and liquid natural gas (LNG) terminals, and oil pipeline transport rates.

13. **Energy Policy Act of 2005** does everything from extending daylight saving time to authorizing $50 million in grants for biomass energy projects to changing the depreciation allowances for gas distribution lines.
14. **State Regulations.** State governments are increasingly active in the environmental and safety areas.

DID YOU KNOW?

The costs of meeting mandated and self-imposed environmental and security requirements are large and continue to grow. Indeed, an increasing portion of new P&E investment is for environmental/security improvement purposes rather than to improve productivity or increase output.

THE BOTTOM LINE

Pre-9/11—Technical innovations and developments in digital information and telecommunications dramatically increased interdependencies among the nation's critical infrastructures. Each infrastructure depends on other infrastructures to function successfully. Disruption in a single infrastructure can generate disturbances within other infrastructures and over long distances, and the pattern of interconnections can extend or amplify the effects of a disruption. The energy infrastructure provides essential fuel to all of the other critical infrastructures and in turn depends on the nation's transportation, communications, finance, and government infrastructures. For example, coal shipments are highly dependent on rail. There are also interdependencies within the energy infrastructure itself, particularly the dependence of petroleum refineries and pipeline pumping stations on a reliable electricity supply and backup generators and utility maintenance vehicles to be supplied with diesel and gasoline fuel.

Post-9/11—The tragic events of 9/11 were a wake-up call for all of us. Like never before in the modern era, the terrorists focused our attention like a laser on security of the homeland and in particular our critical infrastructure. History shows that over the years terrorists have used energy infrastructure as primary targets, causing governments to take extraordinary prevention actions.

REFERENCES AND RECOMMENDED READING

Bond, C. 2002. Statement on S. 2579. *Congressional Record*, daily edition.

Department of Energy. 2009. *Science and technology*. www.energy.gov/sciencetech/index.htm (accessed April 25, 2009).

Africa, it is apparent that energy infrastructure is a preferred target of the terrorists. Moreover, al Qaeda has made it clear that it has an interest in attacking the American pipeline system, a critical component of energy infrastructure.

The obvious question might be: What makes the energy sector such an attractive terrorism target? The energy sector is an attractive target because of the following:

- Energy sector components or assets are spread throughout the nation with little definition of boundaries.
- Much of the energy sector is equipment is one-of-a-kind large machinery that is difficult to replace in the short term.
- Several energy sector components such as substations and remote generating plants were designed and constructed without concern for terrorist intrusion or destructive activities.
- Transmission towers and distribution components are open and widely dispersed.
- Many energy systems are monitored and operated using underprotected SCADA systems.
- As with many industries attempting to economize, many energy sector utilities assign responsibility for safety and security as a collateral duty to a line employee instead of employing a full-time certified safety and security professional (Sullivant 2007).

Even though it was first considered an environmental issue (e.g., petroleum or petroleum-based material spill contaminating the environmental media: air, water, and/or soil), energy sector safety and security has been an issue of congressional interest for many years. Review of the historical incidents listed in sidebar 4.1 provides some perspective on the types of targets and magnitude of the consequences that could result from terrorist attacks on energy sector components. Obviously, these incidents (and others) increase concern over the potential for shutdown of our economy, damage to or shutdown of interrelated critical infrastructure, and the associated potential safety and health impact on surrounding areas. Thus, it is important to ensure that the person designated as site safety/security professional be fully trained in the profession, experienced, and properly certified.

Sidebar 4.1. U.S. Energy Sector Terrorism/Criminal Incidents*

Event 1: Copper-wire theft causes power disruptions (May 20, 2006)—Thieves broke into all four substations of Bangor Hydro Electric Company, breaking locks and tampering with wiring, making the electrical system unsafe in

*The information is a consolidated listing of events, activities, and news accounts as reported by the U.S. Department of Homeland Security and other government agencies. Many of these events were reported by various newspapers throughout the U.S.

addition to the cable theft. Power loss was experienced throughout the area for about six hours.

Event 2: Attempted copper-wire theft (March 29, 2005)—Someone broke into the Omaha Public Power District substation and cut two 13,800-volt power lines. One of the power lines was a backup circuit for part of downtown Omaha's power. Omaha police believed the person who cut the lines was seriously burned. When power employees arrived on the scene, they found a saw and smelled burned flesh. Authorities believe the person who cut the line may have been one of the homeless persons who frequently visit the area of the substation and may have been trying to steal the lines for the copper that's inside.

Event 3: Drug addicts stealing power lines to feed their methamphetamine habits (March 16, 2005)—In Portland, Oregon, police arrested four people with six hundred to eight hundred pounds of stolen power lines. Detectives say the suspects took it from a Portland General Electric (PGE) substation and planned to sell it for scrap and use the money for methamphetamine. PGE says the value of the copper used to make power lines has recently skyrocketed, so thieves are desperate to steal it. Some even risk their lives by trying to swipe live lines. It is illegal for scrap yards to knowingly accept stolen metal. However, there is no law requiring yards o ask where the seller acquired the metal.

Event 4: Criminals hack into power grid (March 11, 2005)—Hackers caused no serious damage to Washington, DC, systems that feed the nation's power grid, but their untiring efforts have heightened concerns that electric companies have failed to adequately fortify defenses against a potential catastrophic strike. The fear is that in a worst-case scenario, terrorists or others could engineer an attack that sets off a widespread blackout and damages power plants, prolonging an outage. The Federal Energy Regulatory Commission and others in the industry said the companies' computer security is uneven. The biggest threat to the grid, analysts said, may come from power companies using older equipment that is more susceptible to attack.

Event 5: Greenpeace activists climb smokestack (February 15, 2005)—Six Greenpeace activists were sentenced to jail terms ranging from five to thirty days for climbing a smokestack at a coal-fire power plant in protest of President Bush's energy policy. The protesters cut a hole in the fence that surrounds Allegheny Energy's Hartfield Ferry Power Station, then climbed the seven-hundred-foot smokestack and unfurled a 2,500-square-foot banner. The six pleaded guilty to misdemeanor charges of reckless endangerment, failure to disperse, disorderly conduct, and defiant trespass.

Event 6: Electricity thefts (January 30, 2005)—Rising energy prices, tighter competition, and the desire to keep bystanders from being electrocuted have led utilities nationwide to fight back with growing fervor. They're hiring full-time investigative teams and running nighttime surveillance. Energy thieves cost U.S. utilities between four billion dollars and six billion dollars yearly. To stem the losses, the nation's investor-owned utility companies hire an average of ten full-time investigators each to root out theft, according to the Edison Electric

Institute, a trade association for U.S shareholder-owned electric companies. Utilities prosecute 10 percent of cases.

Event 7: Gas pipe valve vandalized (January 6, 2005)—In Kearney, Nebraska, vandals turned off a gas pipeline valve to about 150 homes and businesses in frigid temperatures.

Event 8: Electrical power lines sabotaged (December 28, 2004)—In Nevada, eight high-voltage transmission lines servicing the northern region and Reno area were sabotaged. Officials reported the collapse of any one tower could possibly bring down a string of other towers.

Event 9: Attempted bombing of power substation (December 16, 2004)—Utility workers investigating a power outage found cut fencing and an unexploded homemade bomb in a Northern Indiana Public Service Company substation.

Event 10: Vandalism of transmission towers (October 14, 2004)—Four men were reported taking photos and shooting videos outside the headquarters of a Wisconsin transmission company. At the same time, two of the transmission towers collapsed after the bolts were removed.

Event 11: Fake bomb found at electrical tower (October 11, 2004)—A fake bomb at a Philadelphia electrical tower forced the closure of a major highway connecting Philadelphia with its western suburbs for several hours. A box made to look like an explosive but later determined to be a hoax was discovered at the foot of an electric transmission tower. An FBI spokesperson did not know whether the incident was related to the sabotage of an electrical transmission tower in Wisconsin on October 10.

Event 12: Pipeline explosion (September 28, 2004)—Police in New Caney, Texas, connected a pipeline explosion to an August vandalism case. Evidence pointed to one or more suspects behind the explosion and the ensuing six-hour fire. No one was injured, and nearby houses and other structures were not affected. Damage to construction equipment, electricity transmission lines, and the six-inch pipeline itself were estimated to be over one million dollars. Evidence showed that a track hoe and a bucket truck at the explosion site were tampered with.

HOMELAND SECURITY DIRECTIVES

As a result of 9/11, the Homeland Security Department was formed. On matters pertaining to homeland security, Homeland Security Presidential Directives are issued by the president. Each directive has specific meaning and purpose and is carried out by the U.S. Department of Homeland Security. Each directive is listed and summarized in sidebar 4.2.

Sidebar 4.2. Homeland Security Presidential Directives

HSPD-1: Organization and Operation of the Homeland Security Council. (White House) Ensures coordination of all homeland security–related activities among executive departments and agencies and promotes the effective development and implementation of all homeland security policies.

HSPD-2: Combating Terrorism through Immigration Policies. (White House) Provides for the creation of a task force that will work aggressively to prevent aliens who engage in or support terrorist activity from entering the United States and to detain, prosecute, or deport any such aliens who are within the Untied States.

HSPD-3: Homeland Security Advisory System. (White House) Establishes a comprehensive and effective means to disseminate information regarding the risk of terrorist acts to federal, state, and local authorities and to the American people.

HSPD-4: National Strategy to Combat Weapons of Mass Destruction. Applies new technologies and increased emphasis on intelligence collection and analysis, strengthens alliance relationships, and establishes new partnerships with former adversaries to counter this threat in all of its dimensions.

HSPD-5: Management of Domestic Incidents. (White House) Enhances the ability of the United States to manage domestic incidents by establishing a single, comprehensive national incident management system.

HSPD-6: Integration and Use of Screening Information. (White House) Provides for the establishment of the Terrorist Threat Integration Center.

HSPD-7: Critical Infrastructure Identification, Prioritization, and Protection. (White House) Establishes a national policy for federal departments and agencies to identify and prioritize United States critical infrastructure and key resources and to protect them from terrorist attacks.

HSPD-8: National Preparedness. (White House) Identifies steps for improved coordination in response to incidents. This directive describes the way federal departments and agencies will prepare for such a response, including prevention activities during the early stages of a terrorism incident. This directive is a companion to HSPD-5.

HSPD-8 Annex 1: National Planning. Further enhances the preparedness of the United States by formally establishing a standard and comprehensive approach to national planning.

HSPD-9: Defense of United States Agriculture and Food. (White House) Establishes a national policy to defend the agriculture and food system against terrorist attacks, major disasters, and other emergencies.

HSPD-10: Biodefense for the 21st Century. (White House) Provides a comprehensive framework for our nation's biodefense.

HSPD-11: Comprehensive Terrorist-Related Screening Procedures. (White House) Implements a coordinated and comprehensive approach to terrorist-related screening that supports homeland security at home and abroad. This directive builds upon HSPD-6.

HSPD-12: Policy for a Common Identification Standard for Federal Employees and Contractors. (White House) Establishes a mandatory government-wide standard for secure and reliable forms of identification issued by the federal government to its employees and contractors (including contract employees).

HSPD-13: Maritime Security Policy. Establishes policy guidelines to enhance national and homeland security by protecting U.S. maritime interests.

HSPD-16: Aviation Strategy. Details a strategic vision of aviation security while recognizing ongoing efforts, and directs the production of a National Strategy for Aviation Security and supporting plans.

HSPD-18: Medical Countermeasures against Weapons of Mass Destruction. (White House) Establishes policy guidelines to draw upon the considerable potential of the scientific community in the public and private sectors to address medical countermeasure requirements relating to CBRN (chemical, biological, radiological, and nuclear) threats.

HSPD-19: Combating Terrorist Use of Explosives in the United States. (White House) Establishes a national policy and calls for the development of a national strategy and implementation plan on the prevention and detection f, protection against, and response to terrorist use of explosives in the United States.

HSPD-20: National Continuity Policy. (White House) Establishes a comprehensive national policy on the continuity of federal government structures and operations and a single national continuity coordinator responsible for coordinating the development and implementation of federal continuity policies.

HSPD-21: Public Health and Medical Preparedness. (White House) Establishes a national strategy that will enable a level of public health and medical preparedness sufficient to address a range of possible disasters.

Source: Department of Homeland Security (2008).

REFERENCES AND RECOMMENDED READING

CBS News. 2006. *The explosion at Texas City.* www.cbsnews.com/stories/2006/10/26/60 minutes/main2126509.shtml (accessed April 27, 2008).

CR. 2008. Motiva Enterprises settles suit resulting from explosion at Delaware City refinery. Capitol Reports. www.caprep.com/0905038.htm (accessed April 27, 2008).

Crayton, J. W. 1983. Terrorism and the psychology of the self. In *Perspectives on terrorism,* ed. Lawrence Zelic Freedman and Yonah Alexander, 33–41. Wilmington, DE: Scholarly Resources.

Erikson, E. 1994. *Identity of the life cycle.* New York: W. W. Norton.

Ferracuti, F. 1982. A sociopsychiatric interpretation of terrorism. *Annals of the American Academy of Political and Social Science* 463: 129–41.

Fields, R. M. 1979. Child terror victims and adult terrorists. *Journal of Psychohistory* 7(1).

Federal Register. 2007. *Federal register* 17688–17745.

Gurr, T. R. 1971. *Why men rebel.* Princeton, NJ: Princeton University Press.

BusinessWire 2002. *Explosion at First Chemical Corporation plant at Pascagoula, MS.* www.highbeam.com/doc/1G1-92783471.html (accessed April 28, 2008).

HSE 2008-a. *Release of hydrofluoric acid from Marathon Petroleum Refinery, Texas, USA.* U.K. Health and Safety Executive. www.hse.gov.uk/comah/sragtech/casemarathon87.htm (accessed April 27, 2008).

HSE. 2008-b. *PEMEX LPG Terminal, Mexico City, Mexico.* U.K. Health and Safety Executive. www.hse.gov.uk/comah/sragtech/casepemex84.htm (accessed April 27, 2008).

Hudson, R. A. 1999. *The sociology and psychology of terrorism: Who becomes a terrorist and why?* Washington, DC: Library of Congress.

Lees, Frank. 1996. *Loss prevention in the process industries.* Vol. 3, A5.1–A5.11. New York: Butterworth-Heinemann.

Local 1259. 2008. *The Texas City disaster.* www.local1259iaff.org/disaster.html (accessed April 27, 2008).

Long, D. E. 1990. *The anatomy of terrorism.* New York: Free Press.

Margolin, J. 1977. Psychological perspectives on terrorism. In *Terrorism: Interdisciplinary perspectives,* ed. Y. Alexander and S. M. Finger. New York: John Jay Press.

Olson, M. 1971. *The logic of collective action.* Cambridge, MA: Harvard University Press.

Office of Management and Budget. 1998. *Federal conformity assessment activities, circular A-119.* Washington, DC: White House.

Pearlstein, R. 1991. *The mind of the political terrorist.* Wilmington, Delaware: Scholarly Resources.

Sullivant, J. 2007. *Strategies for protecting national critical infrastructure assets: A focus on problem-solving.* New York: Wiley.

U.S. Environmental Protection Agency. 2008. Ashland oil spill. www.epa.gov/reg3hwmd/super/PA/ashlandoil/.

U.S. Fire Administration. 1989. *Phillips Petroleum chemical plant explosion and fire, Pasadena, Texas.* U.S. Fire Administration-TR-035.

Wilkinson, P. 1974. *Political terrorism.* London: Macmillan.

5

Vulnerability Assessment (VA)

Wells Dam, Columbia River, Washington

"In God we trust; all others we monitor."

—*Intercept operator's motto*

Vulnerability means different things to different people . . . many associate vulnerability with a specific set of human activities. . . . (Foster 1987)

INTRODUCTION

One consequence of the events of 9/11 was Department of Homeland Security's (DHS) directive to establish a critical infrastructure protection task force to ensure that activities to protect and secure vital infrastructure are comprehensive and carried out expeditiously. Another consequence is a heightened concern among citizens in the United States over the security of their energy infrastructure (i.e., the uninterrupted supply of electrical power and fuel to power vehicles and homes). As mentioned, along with other critical infrastructure, the energy sector is classified as "vulnerable" in the sense that inherent weaknesses in its operating environment could be exploited to cause harm to the system. There is also the possibility of a cascading effect—a chain of events—due to a terrorist act affecting an energy sector provider, which causes corresponding damage (collateral damage) to other nearby users. In addition to significant damage to the nation's energy sector, entities using energy to produce finished products or provide critical life-saving services (e.g., medical centers and hospitals) that come under terrorist attack could result in: loss of life; shutdown of industry; catastrophic environmental damage to rivers, lakes, and wetlands; contamination of drinking water supplies; other long-term public health impacts; undermining of public confidence in government; and disruption to commerce, the economy, and our normal way of life.

VULNERABILITIES

For the purpose of this text and according to FEMA (2008), vulnerability is defined as any weakness that can be exploited by an aggressor to make an asset susceptible to hazard damage.

In addition, according to the Department of Homeland Security (2009), vulnerabilities are physical features or operational attributes that render an entity open to exploitation or susceptible to a given hazard. Vulnerabilities may be associated with physical (e.g., a broken fence), cyber (e.g., lack of a firewall), or human (e.g., untrained guards) factors.

The FBI (2009) reports that since 9/11 there have been a variety of threats suggesting that U.S. energy facilities are being targeted for terrorist attacks. Although the information often is fragmentary and offers little insight into the timing and mode of an attack, the October 2002 operation against the French supertanker *Limburg* suggests that al Qaeda is serious about hitting the energy sector and its support structure.

In the U.S., al Qaeda appears to believe that an attack on oil and gas structures could do great damage to the U.S. economy. Moreover, terrorist planners probably perceive infrastructure such as dams and power lines as having softer defenses than other facilities.

VULNERABILITY ASSESSMENT (VA)*

A *vulnerability assessment* involves an in-depth analysis of the facility's functions, systems, and site characteristics to identify facility weaknesses and lack of redundancy and determine mitigations or corrective actions that can be designed and implemented to reduce the vulnerabilities. A vulnerability assessment can be a stand-alone process or part of a full risk assessment. During this assessment, the analysis of site assets is based

*Much of the information in this section is from U.S. Department of Energy, *Vulnerability Assessment Methodology: Electric Power Infrastructure* (Washington, DC 2002).

on: (a) the identified threat; (b) the criticality of the assets; and (c) the level of protection chosen (i.e., based on willingness or unwillingness to accept risk).

It is important to point out that post-9/11, all sectors have taken great strides to protect their critical infrastructure. For instance, government and industry have developed vulnerability assessment methodologies for critical infrastructure systems and trained thousands of auditors and others to conduct them.

The actual complexity of vulnerability assessments will range based upon the design and operation of the energy system. The nature and extent of the VA will differ among systems based on a number of factors, including system size, potential population, and safety evaluations. VAs also vary based on knowledge and types of threats, available security technologies, and applicable local, state, and federal regulations. Preferably, a VA is "performance based," meaning that it evaluates the risk to the energy facility based on the effectiveness (performance) of existing and planned measures to counteract adversarial actions. According to the U.S. Environmental Protection Agency (2002), the common elements of energy industry vulnerability assessments are as follows:

- Characterization of the energy sector, including its mission and objectives
- Identification and prioritization of adverse consequences to avoid
- Determination of critical assets that might be subject to malevolent acts that could result in undesired consequences
- Assessment of the likelihood (qualitative probability) of such malevolent acts from adversaries
- Evaluation of existing countermeasures
- Analysis of current risk and development of a prioritized plan for risk reduction

Benefits of Assessments

Energy sector members should routinely perform vulnerability assessments to better understand threats and vulnerabilities, determine acceptable levels of risk, and stimulate action to mitigate identified vulnerabilities. The direct benefits of performing a vulnerability assessment include

- **Build and broaden awareness**—The assessment process directs senior management's attention to security. Security issues, risks, vulnerabilities, mitigation options, and best practices are brought to the surface. Awareness is one of the least expensive and most effective methods for improving the organization's overall security posture.
- **Establish or evaluate against a baseline**—If a baseline has been previously established, an assessment is an opportunity for a checkup to gauge the improvement or deterioration of an organization's security posture. If no previous baseline has been

performed (or the work was not uniform or comprehensive), an assessment is an opportunity to integrate and unify previous efforts, define common metrics, and establish a definitive baseline. The baseline also can be compared against best practices to provide perspective on an organization's security posture.

- **Identify vulnerabilities and develop responses**—Generating lists of vulnerabilities and potential responses is usually a core activity and outcome of an assessment. Sometimes, due to budget, time, complexity, and risk considerations, the response selected for many of the vulnerabilities may be inaction, but after completing the assessment process these decisions will be conscious ones, with a documented decision process and item-by-item rationale available for revisiting issues at scheduled intervals. This information can help drive or motivate the development of a risk management process.
- **Categorize key assets and drive the risk management process**—An assessment can be a vehicle for reaching corporatewide consensus on a hierarchy of key assets. This ranking, combined with threat, vulnerability, and risk analysis, is at the heart of any risk management process. For many organizations, the Y2K scare was the first time a companywide inventory and ranking of key assets was attempted. An assessment allows any organization to revisit that list from a broader and more comprehensive perspective.
- **Develop and build internal skills and expertise**—A security assessment, when not implemented in an "audit" mode, can serve as an excellent opportunity to build security skills and expertise within an organization. A well-structured assessment can have elements that serve as a forum for cross-cutting groups to come together and share issues, experiences, and expertise. External assessors can be instructed to emphasize "teaching and collaborating" rather than "evaluating" (the traditional role). Whatever an organization's current level of sophistication, a long-term goal should be to move that organization toward a capability for self-assessment.
- **Promote action**—Although disparate security efforts may be under way in an organization, an assessment can crystallize and focus management attention and resources on solving specific and systemic security problems. Often the people in the trenches are well aware of security issues (and even potential solutions) but are unable to convert their awareness to action. An assessment provides an outlet for their concerns and the potential to surface these issues at appropriate levels (legal, financial, executive) and achieve action. A well-designed and well-executed assessment not only identifies vulnerabilities and makes recommendations, it also gains executive buy-in, identifies key players, and establishes a set of cross-cutting groups that can convert those recommendations into action.
- **Kick off an ongoing security effort**—An assessment can be used as a catalyst to involve people throughout the organization in security issues, build cross-cutting

teams, establish permanent forums and councils, and harness the momentum generated by the assessment to build an ongoing institutional security effort. The assessment can lead to the creation of either an actual or a virtual (matrixed) security organization.

Vulnerability Assessment Process

Table 5.1 provides an overview of the elements included in the assessment methodology. The elements included in this overview are based on actual in-field experience and lessons learned.

Table 5.1. Basic Elements in Vulnerability Assessments

Element	*Points to Consider*
1. Characterization of the energy entity, including its mission and objectives.	■ What are the important missions of the system to be assessed? Define the highest priority services provided by the utility. Identify the industry's customers: ■ General public ■ Government ■ Military ■ Industrial ■ Critical care ■ Retail operations ■ Firefighting ■ What are the most important facilities, processes, and assets of the system for achieving the mission objectives and avoiding undesired consequences? Describe the ■ Industry facilities ■ Operating procedures ■ Management practices that are necessary to achieve the mission objectives ■ Way the industry operates ■ Treatment processes ■ Storage methods and capacity ■ Energy use and storage ■ Distribution system In assessing those assets that are critical, consider critical customers, dependence on other infrastructures (e.g., chemical, transportation, communications), contractual obligations, single points of failure, chemical hazards, and other aspects of the industry's operations, or availability of industry utilities that may increase or decrease the criticality of specific facilities, processes, and assets.
2. Identification and prioritization of adverse consequences to avoid.	■ Take into account the impacts that could substantially disrupt the ability of the system

(continued)

Table 5.1. ***(continued)***

Element	*Points to Consider*
	to provide a safe and reliable supply of energy. Energy sector systems should use the vulnerability assessment process to determine how to reduce risk associated with the consequences of significant concern. ▪ Ranges of consequences or impacts for each of these events should be identified and defined. Factors to be considered in assessing the consequences may include ▪ Magnitude of service disruption ▪ Economic impact (such as replacement and installation costs for damaged critical assets or loss of revenue due to service outage) ▪ Number of illnesses or deaths resulting from an event ▪ Impact on public confidence in the energy supply ▪ Chronic problems arising from specific events ▪ Other indicators of the impact of each event as determined by the energy sector. Risk reduction recommendations at the conclusion of the vulnerability assessment strive to prevent or reduce each of these consequences.
3. Determination of critical assets that might be subject to malevolent acts that could result in undesired consequences.	▪ What are the malevolent acts that could reasonably cause undesired consequences? ▪ Electronic, computer, or other automated systems that are utilized by the energy sector entities (e.g., supervisory control and data acquisition (SCADA) ▪ The use, storage, or handling of various energy materials (oil, coal, gas, etc.) ▪ The operation and maintenance of such systems
4. Assessment of the likelihood (qualitative probability) of such malevolent acts from adversaries (e.g., terrorists, vandals).	▪ Determine the possible modes of attack that might result in consequences of significant concern based on critical assets of the energy sector entity. The objective of this step of the assessment is to move beyond what is merely possible and determine the likelihood of a particular attack scenario. This is a very difficult task as there is often insufficient information to determine the likelihood of a particular event with any degree of certainty. ▪ The threats (the kind of adversary and the mode of attack) selected for consideration during a vulnerability assessment will dictate, to a great extent, the risk reduction measures that should be designed to counter the threat(s). Some vulnerability assessment methodologies refer to this as a "Design Basis Threat" (DBT), where the threat serves as

Element	Points to Consider
	the basis for the design of countermeasures as well as the benchmark against which vulnerabilities are assessed. It should be noted that there is no single DBT or threat profile for all energy systems in the United States. Differences in geographic location, size of the utility, previous attacks in the local area, and many other factors will influence the threat(s) that the energy sector entity should consider in its assessments. Energy sector entities should consult with the local FBI and/or other law enforcement agencies, public officials, and others to determine the threats upon which their risk reduction measures should be based.
5. Evaluation of existing countermeasures. (Depending on countermeasures already in place, some critical assets may already be sufficiently protected. This step will aid in identification of the areas of greatest concern and help to focus priorities for risk reduction.)	■ What capabilities does the system currently employ for detection, delay, and response? ■ Identify and evaluate current detection capabilities such as intrusion detection systems, energy quality monitoring, operational alarms, guard post orders, and employee security awareness programs. ■ Identify current delay mechanisms such as locks and key control, fencing, structural integrity of critical assets, and vehicle access checkpoints. ■ Identify existing policies and procedures for evaluation and response to intrusion and system malfunction alarms and cyber system intrusions. **It is important to determine the performance characteristics. Poorly operated and maintained security technologies provide little or no protection.** ■ What cyber protection system features does the facility have in place? Assess what protective measures are in place for the SCADA and business related computer information systems such as ■ Firewalls ■ Modem access ■ Internet and other external connections, including wireless data and voice communications ■ Security polices and protocols **It is important to identify whether vendors have access rights and/or "back doors" to conduct system diagnostics remotely.** ■ What security policies and procedures exist, and what is the compliance record for them? Identify existing policies and procedures concerning: ■ Personal security ■ Physical security

(continued)

Table 5.1. ***(continued)***

Element	*Points to Consider*
	▪ Key and access badge control ▪ Control of system configuration and operational data ▪ Chemical and other vendor deliveries ▪ Security training and exercise records

6. Analysis of current risk and development of a prioritized plan for risk reduction.	▪ Information gathered on threat, critical assets, energy sector operations, consequences, and existing countermeasures should be analyzed to determine the current level of risk. The utility should then determine whether current risks are acceptable or risk reduction measures should be pursued.
	▪ Recommended actions should measurably reduce risks by reducing vulnerabilities and/or consequences through improved deterrence, delay, detection, and/or response capabilities or by improving operational policies or procedures. Selection of specific risk reduction actions should be completed prior to considering the cost of the recommended action(s). Facilities should carefully consider both short- and long-term solutions. An analysis of the cost of short- and long-term risk reduction actions may impact which actions the utility chooses to achieve its security goals.
	▪ Facilities may also want to consider security improvements. Security and general infrastructure may provide significant multiple benefits. For example, improved treatment processes or system redundancies can both reduce vulnerabilities and enhance day-to-day operation.
	▪ Generally, strategies for reducing vulnerabilities fall into three broad categories:
	  ▪ Sound business practices—affect policies, procedures, and training to improve the overall security-related culture at the chemical facility. For example, it is important to ensure rapid communication capabilities exist between public health authorities and local law enforcement and emergency responders.
	  ▪ System upgrades—include changes in operations, equipment, processes, or infrastructure itself that make the system fundamentally safer.
	  ▪ Security upgrades—improve capabilities capabilities for detection, delay, or response.

In table 5.1, step 3 deals with identification of asset criticality. This is an important first step in any vulnerability assessment. Identifying asset criticality serves several functions:

- It enables more careful consideration of factors that affect risk, including threats, vulnerabilities, and consequences of loss or compromise of the asset.
- It enables more focused and thorough consideration of loss or compromise of the asset.
- It enables leaders to develop robust methods for managing consequences of asset loss (restoration).
- It provides a means to increase awareness of a broad range of employees to protect truly critical assets and to differentiate in policies and procedures the heightened protection they require.

As previously indicated, identifying the criticality of assets is used primarily to focus the vulnerability analysis efforts. It also assists with the ranking of various recommendations for reducing vulnerabilities. Let's take a look at what electrical power infrastructure critical assets might include:

Physical

- Generators
- Substations
- Transformers
- Transmission lines
- Distribution lines
- Control center
- Warehouses
- Office buildings
- Internal and external infrastructure dependencies

Cyber

- SCADA systems
- Networks
- Databases
- Business systems
- Telecommunications

Interdependencies

- Single-point nodes of failures
- Critical infrastructure components of high reliance

VULNERABILITY ASSESSMENT METHODOLOGY (U.S. DEPARTMENT OF ENERGY 2002)

Vulnerability assessment methodology consists of ten elements. Each element along with a description is listed below.

1. Network architecture
2. Threat environment
3. Penetration testing
4. Physical security
5. Physical asset analysis
6. Operations security
7. Policies and procedures
8. Impact analysis
9. Infrastructure interdependencies

10 Risk characterization

Network Architecture

This element provides an analysis of the information assurance features of the information network(s) associated with the organization's critical information systems. Information examined should include network topology and connectivity (including subnets), principal information assets, interface and communication protocols, function and linage of major software and hardware components (especially those associated with information security such as intrusion detectors), and policies and procedures that govern security features of the network.

Procedures for information assurance in the system, including authentication of access and management of access authorization, should be reviewed. The assessment should identify any obvious concerns related to architectural vulnerabilities, as well as operating procedures. Existing security plans should be evaluated, and the results of any prior testing should be analyzed. Results from the network architecture assessment should include potential recommendations for changes in the information architecture, functional areas and categories where testing is needed, and suggestions regarding system design that would enable more effective information and information system protection.

Three techniques are often used in conducting the network architecture assessment:

1. Analysis of network and system documentation during and after the site visit
2. Interview with facility staff, managers, and chief information officer
3. Tours and physical inspections of key facilities

Threat Environment

Development of a clear understanding of the threat environment is a fundamental element of risk management. When combined with an appreciation of the value of the information assets and systems and the impact of unauthorized access and subsequent malicious activity, an understanding of threats provides a basis for better defining the level of investment needed to prevent such access.

The threat of a terrorist attack to the energy sector infrastructure is real and could come from several areas, including physical, cyber, and interdependency. In addition, threats could come from individuals or organizations motivated by financial gain or persons who derive pleasure from such penetration (e.g., recreational hackers, disgruntled employees). Other possible sources of threats are those who want to accomplish extremist goals (e.g., environmental terrorists, antinuclear advocates) or embarrass one or more organizations.

This element should include a characterization of these and other threats, identification of trends in these threats, and ways in which vulnerabilities are exploited. To the extent possible, characterization of the threat environment should be localized, that is, within the organization's service area.

Penetration Testing

The purpose of network penetration testing is to utilize active scanning and penetration tools to identity vulnerabilities that a determined adversary could easily exploit. Penetration testing can be customized to meet the specific needs and concerns of the energy sector unit or utility. In general, penetration testing should include a test plan and details on the rules of engagement (ROE). It should also include a general characterization of the access points to the critical information systems and include a general characterization of the access points to the critical information systems and communication interface connections, modem network connections, access points to principal network routers, and other external connections. Finally, penetration testing should include identified vulnerabilities and, in particular, whether access could be gained to the control network or specific subsystem or devices that have a critical role in assuring continuity of service.

Penetration testing consists of an overall process of establishing the ground rules or ROE for the test; establishing a white cell for continuous communication; developing a format or methodology for the test; conducting the test; and generating a final report that details methods, findings, and recommendations.

Penetration testing methodology consists of three phases: reconnaissance, scenario development, and exploitation. A one-time penetration test can provide the utility with valuable feedback; however, it is far more effective if performed on a regular basis. Repeated testing is recommended because new threats develop continuously, and the networks, computers, and architecture of the energy sector unit or utility are likely to change over time.

Physical Security

The purpose of physical security assessment is to examine and evaluate the systems in place (or being planned) and to identify potential improvements in this area for the sites evaluated. Physical security systems include access controls, barriers, locks and keys, badges and passes, intrusion detection devices and associated alarm reporting and display, closed-circuit television (assessment and surveillance), communications equipment (telephone, two-way radio, intercom, cellular), lighting (interior and exterior), power sources (line, battery, generator), inventory control, postings (signs), security system wiring, and protective force (see chapter 9 for greater detail on security devices). Physical security systems are reviewed for design, installation operation, maintenance, and testing.

The physical security assessment should focus on those sites directly related to the critical facilities, including information systems and assets required for operation. Typically included are facilities that house critical equipment or information assets or networks dedicated to the operation of electricity, oil, or gas transmission, storage, or delivery systems. Other facilities can be included on the basis of criteria specified by the organization being assessed. Appropriate levels of physical security are contingent upon the value of company assets, the potential threats to these assets, and the cost associated with protecting the assets. Once the cost of implementing/maintaining physical security programs is known, it can be compared to the value of the company assets, thus providing the necessary information for risk management decisions. The focus of the physical security assessment task is determined by prioritizing the company assets, that is, the most critical assets receive the majority of the assessment activity.

At the start of the assessment, survey personnel should develop a prioritized listing of company assets. This list should be discussed with company personnel to identify areas of security strengths and weaknesses. During these initial interviews, assessment areas that would provide the most benefit to the company should be identified; once known, they should become the major focus of the assessment activities.

The physical security assessment of each focus area usually consists of the following:

- Physical security program (general)
- Physical security program (planning)
- Barriers
- Access controls/badges
- Locks/keys
- Intrusion detection systems
- Communications equipment
- Protective force/local law enforcement agency

The key to reviewing the above topics is not to just identify whether they exist but to determine the appropriate level that is necessary and consistent with the value of the asset being protected. The physical security assessment worksheets provide guidance on appropriate levels of protection.

Once the focus and content of the assessment task have been identified, the approach to conducting the assessment can be either at the "implementation level" or at the "organizational level." The approach taken depends on the maturity of the security program.

For example, a company with a solid security infrastructure (staffing plans/procedures, funding) should receive a cursory review of these items; however, facilities where the security programs are being implemented should receive a detailed review. The security staff can act upon deficiencies found at the facilities, once reported.

For companies with an insufficient security organization, the majority of time spent on the assessment should take place at the organizational level to identify the appropriate staffing/funding necessary to implement security programs to protect company assets. Research into specific facility deficiencies should be limited to finding just enough examples to support any staffing/funding recommendations.

Physical Asset Analysis

The purpose of the physical asset analysis is to examine the systems and physical operational assets to ascertain whether vulnerabilities exist. Included in this element is an examination of asset utilization, system redundancies, and emergency operating procedures. Consideration should also be given to the topology and operating practices for electricity and gas transmission, processing, storage, and delivery, looking specifically for those elements that either singly or in concert with other factors provide a high potential for disrupting service. This portion of the assessment determines company and industry trends regarding these physical assets. Historic trends, such as asset utilization, maintenance, new infrastructure investments, spare parts, SCADA linkages, and field personnel, are part of the scoping element.

The proposed methodology for physical assets is based on a macro-level approach. The analysis can be performed with company data, public data, or both. Some companies might not have readily available data or might be reluctant to share that data.

Key output from analysis should be graphs that show trends. The historic data analysis should be supplemented with on-site interviews and visits. Items to focus on during a site visit include the following:

- Trends in field testing
- Trends in maintenance expenditures
- Trends in infrastructure investments
- Historic infrastructure outages
- Critical system components and potential system bottlenecks
- Overall system operation controls
- Use and dependency of SCADA systems
- Linkages of operation staff with physical and IT security
- Adequate policies and procedures
- Communications with other regional utilities
- Communications with external infrastructure providers
- Adequate organizational structure

Operations Security

Operations security (OPSEC) is the systematic process of denying potential adversaries (including competitors or their agents) information about capabilities and intentions of the host organization. OPSEC involves identifying, controlling, and protecting generally nonsensitive activities concerning planning and execution of sensitive activities. The OPSEC assessment reviews the processes and practices employed for denying adversary access to sensitive and nonsensitive information that might inappropriately aid or abet an individual's or organization's disproportionate influence over system operation (e.g., electric markets, grid operations). This assessment should include a review of security training and awareness programs, discussions with key staff, and tours of appropriate principal facilities. Information that might be available through public access should also be reviewed.

Policies and Procedures

The policies and procedures by which security is administered (1) provide the basis for identifying and resolving issues; (2) establish the standards of reference for policy implementation; and (3) define and communicate roles, responsibilities, authorities, and accountabilities for all individuals' and organizations' interface with critical systems. They are the backbone for decisions and day-to-day security operations. Security policies and procedures become particularly important at times when multiple parties must interact to effect a desired level of security and when substantial legal ramifications could result from policy violations. Policies and procedures should be reviewed to determine whether they (1) address the key factors affecting security; (2) enable effective compliance, implementation, and enforcement; (3) reference or conform to established standards; (4) provide clear and comprehensive guidance; and (5) effectively address the roles, responsibilities, accountabilities, and authorities.

The objective of the policies and procedures assessment task is to develop a comprehensive understanding of how a facility protects its critical assets through the development and implementation of policies and procedures. Understanding and assessing this area provide a means of identifying strengths and areas for improvements that can be achieved through

- Modification of current policies and procedures
- Implementation of current policies and procedures
- Development and implementation of new policies and procedures
- Assurance of compliance with policies and procedures
- Cancellation of policies and procedures that are no longer relevant or are inappropriate for the facility's current strategy and operations

Impact Analysis

A detailed analysis should be conducted to determine the influence that exploitation of unauthorized access to critical facilities or information systems might have on an organization's operations (e.g., market and/or physical operations). In general, such an analysis would require a thorough understanding of (1) the applications and their information processing, (2) decisions influenced by this information, (3) independent checks and balances that might exist regarding information upon which decisions are made, (4) factors that might mitigate the impact of unauthorized access, and (5) secondary impacts of such access (e.g., potential destabilization of organizations serving the grid, particularly those affecting reliability or safety). Similarly, the physical chain of events following disruption, including the primary, secondary, and tertiary impacts of disruption, should be examined.

The purpose of the impact analysis is to help estimate the impact that outages could have on an energy sector unit. Outages in electric power, natural gas, and oil can have significant financial and external consequences to an energy sector unit. The impact analysis provides an introduction to risk characterization by providing quantitative estimates of these impacts so that the energy sector unit can implement a risk management program and weigh the risks and costs of various mitigation measures.

Infrastructure Interdependencies

The term *infrastructure interdependencies* refers to the physical and electronic (cyber) linkages within and among our nation's critical infrastructures—energy (electrical power, oil, natural gas), telecommunications, transportation, water supply systems, banking and finance, emergency services, and government services. This task identifies the direct infrastructure linkages between and among the infrastructures that support critical facilities as recognized by the organization. Performance of this task requires a detailed understanding of an organization's functions and internal infrastructures and how these link to external infrastructures.

The purpose of the infrastructure interdependencies assessment is to examine and evaluate the infrastructures (internal and external) that support critical facility functions, along with their associated interdependencies and vulnerabilities.

Risk Characterization

Risk characterization provides a framework for prioritizing recommendations across all task areas. The recommendations for each task area are judged against a set of criteria to help prioritize the recommendations and assist the organization in determining the appropriate course of action. It provides a framework for assessing vulnerabilities, threats, and potential impacts (determined in the other tasks). In addition, the existing risk analysis and management process at the organization should be reviewed and, if appropriate, utilized for prioritizing recommendations. The degree to which corporate risk management includes security factors is also evaluated.

VULNERABILITY ASSESSMENT PROCEDURES

Vulnerability assessment procedures can be conducted using various methodologies. For example, the checklist analysis is an effective technology. In addition, Pareto analysis (80/20 principle), relative ranking, preremoval risk assessment (PRRA), change analysis, failure mode and effects analysis (FMEA), fault tree analysis, event tree analysis, what-if analysis, and Hazard and Operability (HAZOP) can be used in conducting the assessment.

Based on our experience, the what-if analysis and HAZOP seem to be the most user-friendly methodologies to use. A sample what-if analysis procedural outline is presented below, followed by a brief explanation of and outline for conducting HAZOP.

What-If Analysis Procedure/Sample What-If Questions

The steps in a what-if checklist analysis are as follows:

1. Select the team (personnel experienced in the process).
2. Assemble information (piping and instrumentation drawings (P&IDs), process flow diagrams (PFDs), operating procedures, equipment drawings, and so on.

3. Develop a list of what-if questions.
4. Assemble your team in a room where each team member can view the information.
5. Ask each what-if question in turn and determine:

- What can cause the deviation from design intent that is expressed by the question?
- What adverse consequences might follow?
- What are the existing design and procedural safeguards?
- Are these safeguards adequate?
- If these safeguards are not adequate, what additional safeguards does the team recommend?

6. As the discussion proceeds, record the answers to these questions in tabular format.
7. Do not restrict yourself to the list of questions that you developed before the project started. The team is free to ask additional questions at any time.
8. When you have finished the what-if questions, proceed to examine the checklist. The purpose of this checklist is to ensure that the team has not forgotten anything. While you are reviewing the checklist, other what-if questions may occur to you.
9. Make sure that you follow up all recommendations and action items that arise from the hazards evaluation.

HAZOP Analysis

The HAZOP analysis technique uses a systematic process to (1) identify possible deviations from normal operations and (2) ensure that safeguards are in place to help prevent accidents. The HAZOP uses special adjectives (such as speed, flow, pressure, etc.) combined with process conditions (such as "more," "less," "no," etc.) to systematically consider all credible deviations from normal conditions. The adjectives, called guide words, are a unique feature of HAZOP analysis.

In this approach, each guide word is combined with relevant process parameters and applied at each point (study node, process section, or operating step) in the process that is being examined.

Guide Words	Meaning
No	Negation of the Design Intent
Less	Quantitative Decrease
More	Quantitative Increase
Part Of	Other Material Present by Intent
As Well As	Other Materials Present Unintentionally

Reverse	Logical Opposite of the Intent
Other Than	Complete Substitution

Common HAZOP Analysis Process Parameters

Flow	Time	Frequency	Mixing
Pressure	Composition	Viscosity	Addition
Temperature	pH	Voltage	Separation
Level	Speed	Information	Reaction

The following is an example of creating deviations using guide words and process parameters:

Guide Words		Parameter		Deviation
NO	+	FLOW	=	NO FLOW
MORE	+	PRESSURE	=	HIGH PRESSURE
AS WELL AS	+	ONE PHASE	=	TWO PHASE
OTHER THAN	+	OPERATION	=	MAINTENANCE
MORE	+	LEVEL	=	HIGH LEVEL

Guide words are applied to both the more general parameters (e.g., react, mix) and the more specific parameters (e.g., pressure, temperature). With the general parameters, it is not unusual to have more than one deviation from the application of one guide word. For example, "more reaction" could mean either that a reaction takes place at a faster rate, or that a greater quantity of product results. On the other hand, some combination of guide words and parameters will yield no sensible deviation (e.g., "as well as" with "pressure").

HAZOP Procedure

1. Select the team.
2. Assemble information (P&IDs, PFDs, operating procedures, equipment drawings, etc.).
3. Assemble your team in a room where each team member can view P&IDs.
4. Divide the system you are reviewing into nodes (you can present the nodes, or the team can choose them as you go along).
5. Apply appropriate deviations to each node. For each deviation, address the following questions:

   - What can cause the deviation from design intent?
   - What adverse consequences might follow?

- What are the existing design and procedural safeguards?
- Are these safeguards adequate?
- If these safeguards are not adequate, what does the team recommend?

6. As the discussion proceeds, record the answers to these questions in tabular format.

VULNERABILITY ASSESSMENT: CHECKLIST PROCEDURE

In performing the vulnerability assessment of any energy sector unit or facility, one of the simplest methodologies to employ is the checklist. In this section we present a basic outline that is not all inclusive but does contain questions related to each of the infrastructure interdependency survey checklists and their subsections. These sample questions are intended for use by the assessment teams during preparation for interviews with facility representatives to help assure that all relevant aspects of the critical infrastructures are considered in the survey.

Electric Power Supply and Distribution Checklist

Primary Source of Electric Power

- If the primary source of electric power is a commercial source, are there multiple independent feeds? If so, describe the feeds and their locations.
- If the primary source of electric power is a system operated by the facility or asset, what type of system is it?
- If a facility-operated primary electric generation system is used, what are the fuel or fuels used?
- If petroleum fuel is used, what quantity of fuel is stored on site for the primary electric generation system, and how long it would last under different operating conditions?
- If the fuel is stored on site, are arrangements and contracts in place for resupply and management of the fuel?

Electric Distribution System

- Are the components of the electric system that are located outside of buildings (such as generators, fuel storage facilities, transformers, transfer switches) protected from vandalism or accidental damage by fences or barriers? If so, describe the type of protection and level of security it provides.
- Are the various sources of electric power and the components of the internal electric distribution system such that they may be isolated for maintenance or replacement without affecting the critical functions of the asset/facility? If not, describe the limitations.

- Have any single points of failure been identified for the electric power supply and distribution system? If so, list them and describe.

Backup Electric Power Systems

- Are there additional emergency sources of electric supply beyond the primary system (such as multiple independent commercial feeds, backup generators, uninterruptible power supply [UPS])? If there are, describe them.
- If there is a central UPS, does it support all the critical functions of the asset/facility in terms of capacity and connectivity? Specify for how long it can operate on battery power, and list any potentially critical functions that are not supported.
- If there is a backup generator system, does it support all the critical functions of the facility in terms of capacity and connectivity? Specify the fuel, and list any potentially critical functions that are not supported.
- Is the fuel for the backup generator system a petroleum fuel? If yes, specify the quantity stored on site and how long it would last.
- If the fuel is stored on site, are arrangements and contracts in place for resupply and management of the fuel?

Commercial Electric Power Sources

- How many substations feed the area of the asset facility and the asset/facility itself? That is, is the area supplied by multiple substations? If more than one, which ones have sufficient individual capacities to supply the critical needs of the asset/facility?
- How many distinct independent transmission lines supply the substations?

Commercial Electric Power Pathways

- Are the power lines into the area of the asset/facility and into the asset/facility itself aboveground (on utility poles), buried, or a combination of both? If both, indicate locations of portions aboveground.
- Do the power lines from these substations follow independent pathways to the area of the asset/facility? If not, specify how often and where they intersect or follow the same corridor.
- Are the paths of the power lines colocated with the rights-of-way of other infrastructures? If yes, indicate how often and where they follow the same rights-of-way and the infrastructures that are colocated.
- Are the paths of the power lines located in areas susceptible to natural or accidental damage (such as overhead lines near highways or power lines across bridges, dams, or landslide areas)? If yes, indicate the locations and types of potential disruptions.

Commercial Electric Power Contracts

- What type of contract does the asset/facility have with the electric power distribution company or transmission companies? Specify the companies involved and whether there is a direct physical link (distribution or transmission power line) to each company.
- If there is an interruptible contract (even in part), what are the general conditions placed on interruptions such as the minimum quantity that is not interruptible, the maximum number of disruptions per time period, and the maximum duration of disruptions? Has electrical service been interrupted in the past? If yes, describe the circumstances and any effect the outages have had on the critical functions and activities of the asset/facility.

Historical Reliability

- Historically, how reliable has commercial electric power been in the area? Quantify in terms of annual number of disruptions and their durations.
- Typically, when power outages occur, are they of significant duration (as opposed to just a few seconds or minutes)? Quantify the duration of the outages.
- Have there ever been electric power outages of sufficient frequency and duration so as to affect the critical functions and activities of the asset/facility?

Petroleum Fuels Supply and Storage Checklist

Uses of Petroleum Fuels

- Are petroleum fuels used in normal operations at the asset/facility? If yes, specify the steps and uses.
- Are petroleum fuels used during contingency or emergency operations such as for backup equipment or repairs? If yes, specify the types of fuels and their uses.

Reception Facilities

- How are the various petroleum fuels normally delivered to the asset/facility? Indicate the delivery mode and normal frequency of shipments of each fuel type.
- Under maximum use–rate conditions, are there sufficient reception facilities (truck racks, rail sidings, surge tank capacity, barge moorings) to keep up with maximum contingency or emergency demand)? If no, explain where the expected shortfalls would be and their impacts.
- Are the petroleum fuel delivery pathways colocated with the rights-of-way of other infrastructures or located in areas susceptible to natural or accidental damage (across bridges or dams, in earthquake or landslide areas)? If yes, indicate the locations and types of potential disruptions.

- Are contingency procedures in place to allow for alternative modes or routes of delivery? If yes, describe these alternatives and indicate whether they have sufficient capacity to fully support the critical functions and activities of the asset/facility.

Supply Contracts

- Are contracts in place for the supply of petroleum fuels? Specify the contractors, the types of contracts, the modes of transport (pipeline, rail car, tank truck), and the frequency of normal shipments.
- Are arrangements for emergency deliveries of petroleum fuels in place? Indicate the basic terms of the contracts in terms of the maximum time to delivery and the minimum and maximum quantity per delivery. Also, indicate whether these terms are such that there may be effects on the critical functions and activities of the asset/facility.

Natural Gas Supply Checklist

Sources of Natural Gas

- How many city gate stations supply the natural gas distribution system in the area of the asset/facility and the asset facility itself? If more than one, which ones are critical to maintaining the distribution system?
- How many distinct independent transmission pipelines supply the city gate stations? Indicate whether an individual gate station is supplied by more than one transmission pipeline and which stations are supplied by independent transmission pipelines.

Pathways of Natural Gas

- Do the distribution pipelines from the individual city gate stations follow independent pathways to the area of the asset/facility? If not, specify how often and where they intersect or follow the same corridor.
- Are the paths of the pipelines colocated with the rights-of-way of other infrastructures? If yes, indicate how often and where they follow the same rights-of-way and the infrastructures that are colocated.
- Are the paths of the pipelines located in areas susceptible to natural or accidental damage (such as across bridges or dams, in earthquake or landslide areas)? If yes, indicate the locations and types of potential disruptions.
- Is the local distribution system well integrated (i.e., can gas readily get from any part of the system to any other part of the system)?

Natural Gas Contracts

- Does the asset/facility have a firm delivery contract, an interruptible contract, or a mixed contract with the natural gas distribution company or the transmission companies? Specify the companies involved and whether there is a direct physical link (pipeline) to each company.

- If there is an uninterruptible contract (even in part), what are the general conditions placed on interruptions such as the minimum quantity that is not interruptible, the maximum number of disruptions per time period, and the maximum duration of disruptions? Has natural gas service been interrupted in the past? If yes, describe the circumstances and any effect the outages have had on the critical functions and activities of the asset/facility.
- Does the asset/facility have storage or some other sort of special contracts with natural gas transmission or storage companies? If yes, briefly describe the effect on sustaining a continuous supply of natural gas to the asset/facility.
- In case of a prolonged disruption of natural gas supply, are contingency procedures in place to allow for the use of alternative fuels (such as on-site propane-air, liquefied petroleum gas, or petroleum fuels)? If yes, describe these alternatives and indicate whether they have sufficient capacity to fully support the critical functions and activities of the asset/facility.

Historical Reliability

- Historically, how reliable has the natural gas supply been in the area? Quantify by describing any unscheduled or unexpected disruptions. Were there any effects on the critical functions and activities of the asset/facility?
- If operating under an uninterruptible service agreement, has natural gas service ever been curtailed? If yes, how often, for how long, and were there any effects on the critical functions and activities of the asset/facility?

REFERENCES AND RECOMMENDED READING

Belke, J. C. 2001. Chemical accident risks in U.S. industry—A preliminary analysis of accident risk data from U.S. hazardous chemical facilities. In *Proceedings of the 10th international symposium on loss prevention and safety promotion in the process industries*, Stockholm, Sweden. Amsterdam: Elsevier Science.

Clark, R. M., and Deininger, R. A. 2000. Protecting the nation's critical infrastructure: The vulnerability of U.S. water supply systems. *J. Contingen. Crisis Management* 8(2): 76–80.

CBO. 2004. *Homeland security and the private sector.* Congressional Budget Office. www.cbo .gov/ftpdocs/60xx/doc6042/12-20-HomelandSecurity.pdf (accessed May 2, 2008).

Department of Homeland Security. 2009. National infrastructure protection plan. www.dhs .gov/ xlibrary/assets/NIPP_Plan.pdf (accessed May 1, 2009).

FBI. 2009. Terrorism threats to energy sector. Congressional testimony of FBI director. www .fbi.gov/congress/congress03/mueller021103.htm (accessed May 5, 2009).

FEMA. 2008. *FEMA452: Risk assessment: A how to guide.* Federal Emergency Management Agency. www.fema.gov/plan/prevent/rms/rmsp452.shtm (accessed May 1, 2008).

Foster, S. S. D. 1987. Fundamental concepts in aquifer vulnerability, pollution risk and protection strategy. In *Vulnerability of soil and groundwater pollutants*, ed. W. van Duijvenbooden and H. G. van Waegeningh. The Hague: TNO Committee on Hydrological Research: Proceedings and Information 38.

Government Accountability Office. 2004. *RMP-covered industrial processes and off-site consequences of worst-case chemical releases.* Washington, DC: U.S. Government Accountability Office, GAO-04-482T Appendix I.

———. 2005. *Wastewater facilities: Experts' view on how federal funds should be spent.* U.S. Government Accountability Office, GAO-05-165.

Minter, J. G. 1996. Prevention of chemical accidents still a challenge. *Occupational hazards*, September.

Spellman, F. R. 1997. *A guide to compliance for PSM/RMP.* Lancaster, PA: Technomic.

U.S. Department of Energy. 2002. *Vulnerability assessment methodology: Electric power infrastructure.* Washington, DC.

U.S. Environmental Protection Agency. 2002. *Vulnerability assessment fact sheet.* EPA 816-F-02-025. www.epa.gov/ogwdw/security/index.html (accessed May 2006).

6

Preparation: When Is Enough Enough?

Security control station, Rocky Reach Dam, Washington

"We must take the battle to the enemy, disrupt his plans, and confront the worst threats before they emerge."

—*George W. Bush*

Governor Tom Ridge points out the security role for public service professionals: "Americans should find comfort in knowing that millions of their fellow citizens are working every day to ensure our security at every level—federal, state, county, municipal. These are dedicated professionals who are good at what they do. I've seen it up close, as Governor of Pennsylvania. . . . But there may be gaps in the system. The job of the Office of Homeland Security will be to identify those gaps and work to close them. Now, obviously, the further removed we get from September 11, I think the natural tendency is to let down our guard. Unfortunately, we cannot do that. The government will continue to do everything we can to find and stop those who seek to harm us. And I believe we owe it to the American people to remind them that they must be vigilant, as well." (Henry 2002)

INTRODUCTION

The possibility of energy terrorism—attacks on the U.S. energy infrastructure—doesn't generate the same attention as potential nuclear, biological, or chemical terrorism. However, because of the seriousness of the threat of terrorism to the nation's energy sector and the enormous economic implications of such attacks, USDOE, USEPA, and other agencies have worked nonstop since 9/11 in gathering and providing as much advice and guidance as possible to aid energy sector personnel in protecting energy sector production facilities and associated critical support infrastructure. In this chapter, we provide an overview of important tools that can be used in protecting energy sector and petrochemical industries (i.e., manufacturers of gasoline, cosmetics, detergents, fertilizers, synthetic fabrics, asphalt, and plastics) against the threat of terrorism. In the discussion, even though a chemical industry sector issue, we include the petrochemical industry because of its dependence on petroleum and natural gas supplies.

THREATS AND INCIDENTS

Based on evidence of potential losses from past accidents, indication of the potential human and environmental losses and economic costs from an attack on a large energy sector facility or producer comes from major accidents that have occurred both abroad and in the United States. Those events indicate that the human and environmental losses could be significant (CBO 2004).

Energy sector threats and incidents may be of particular concern due to the range of potential consequences:

- Creating an adverse impact on public health within a population
- Disrupting system operations and interrupting the supply of energy
- Causing physical damage to system infrastructure
- Reducing public confidence in the energy supply system
- Long-term denial of energy and the cost of replacement

Keep in mind that some of these consequences would only be realized in the event of a successful terrorist incident; however, the mere threat of terrorism can also have an adverse impact on industries that depend on a safe, steady supply of energy. In addition, the economic implications of such attacks are potentially enormous. For example, many believe that the reason we are looking at oil at more than fifty dollars per barrel is the fact that we have a "terror premium" factored into the price of a barrel of oil.

While it is important to consider the range of possibilities associated with a particular threat, assessments are typically based on the probability of a particular occurrence. Determining probability is somewhat subjective and is often based on intelligence and previous incidents. As mentioned, there are historical accounts of accidental incidents that have caused tremendous death and destruction.

Threat Warning Signs

A threat warning is an occurrence or discovery that indicates a potential threat that triggers an evaluation of the threat. It is important to note that these warnings must be evaluated in the context of typical industry activity and previous experience in order to avoid false alarms. Following is a brief description of potential warnings.

- *Security Breach.* Physical security breaches, such as unsecured doors, open hatches, and unlocked/forced gates, are probably the most common threat warnings. In most cases, the security breach is likely related to lax operations or typical criminal activity such as trespassing, vandalism, and theft. However, it may be prudent to assess any security breach with respect to the possibility of attack.
- *Witness Account.* Awareness of an incident may be triggered by a witness account of tampering. Energy sector sites/facilities should be aware that individuals observing suspicious behavior near energy sector facilities/plants will likely call 911 and not the plant. In this case, the incident warning technically might come from law enforcement, as described below. Note: The witness may be a plant employee engaged in normal duties.
- *Direct Notification by Perpetrator.* A threat may be made directly to the energy sector site, plant, or facility, either verbally or in writing. Historical incidents would indicate that verbal threats made over the phone are more likely than written threats. While the notification may be a hoax, threatening an energy sector unit is a crime and should be taken seriously.
- *Notification by Law Enforcement.* An energy sector site/facility may receive notification about a threat directly from law enforcement, including local, county, state, or federal agencies. As discussed previously, such a threat could be a result of suspicious activity reported to law enforcement, either by a perpetrator, a witness, or the news media. Other information, gathered through intelligence or informants, could also lead law enforcement to conclude that there may be a threat to the energy sector site/facility. While law enforcement will have to take the lead in the criminal investigation, the energy sector site/facility has primary responsibility for the safety of it equipment, energy-producing materials and processes, and public health. Thus, the plant's role will likely be to help law enforcement to appreciate the public health implications of a particular threat as well as the technical feasibility of carrying out a particular threat.

- *Notification by News Media.* A threat to destroy an energy site/facility might be delivered to the news media, or the media may discover a threat. A conscientious reporter would immediately report such a threat to the police, and either the reporter or the police would immediately contact the energy sector site/facility. This level of professionalism would provide an opportunity for the plant to work with the media and law enforcement to assess the credibility of the threat before any broader notification is made.
- *Public Health Notification.* In this case, the first indication that energy blackouts or energy sector site/facility emergencies (e.g., sabotage) have occurred is the appearance of victims in local emergency rooms and health clinics. Energy sector sites/facilities may therefore be notified, particularly if the cause is unknown or linked to energy materials (oil, gas, coal, or by-products and/or petrochemicals). An incident triggered by a public health notification is unique in that at least a segment of the population has been exposed to a harmful substance. If this agent is a hazardous petrochemical, then the time between exposure and onset of symptoms may be on the order of hours, and thus there is the potential that the contaminant is still present.

RESPONSE TO THREATS

Note: This section is not designed to discuss what specific steps to take in responding to a terrorist threat. Rather, the questions addressed in this section are "Why is it necessary to plan to respond to energy sector threats at all?" and "When have I done enough?"

Federal, state, and local programs already exist that—with varying degrees of effectiveness—encourage or require the operators of energy sector sites/facilities to boost their efforts to promote safety and security and to share information that can help local governments plan for emergencies (Schierow 2004).

Proper planning is a delicate process because public health measures are rarely noticed or appreciated (like buried utility pipes, they are often hidden functions) except when they fail—then they are very visible. Consumers are particularly upset by unreliable energy supply systems and energy products or practices that produce unsafe (contaminated) environmental media—water, air, and/or soil—because they are often viewed as entitlements, and indeed, it is reasonable for consumers to expect a high-quality, safe environment. Public health failures during response to contamination threats often take the form of too much or too little action. The results of too little action, including no response at all, can have disastrous consequences potentially resulting in public injuries or fatalities. On the other hand, a disproportionate response to fuel oil or other petrochemical contamination threats that have not been corroborated (i.e., determined to be "credible") can also have serious repercussions when, for example, otherwise safe drinking water is unavailable because it has been contaminated with petrochemicals. Not only would the water be unavailable for human consumption, but it would also be unavailable for sanitation, firefighting, industry, and the many other uses of public water supply. Although precise estimates are not available, information on accidents in the energy sector, company assessment of what could happen in a severe release or explosion, and actual terrorist incidents involving fuel supplies and/or petrochemicals suggest that the risk of attack is real and the losses could be significant. These adverse impacts must be considered when evaluating response options to a petrochemical contamination threat.

Considering the potential risks of an inappropriate response to a severe petrochemical release or explosion threat, it is clear that a systematic approach is needed to evaluate petrochemical contamination threats. One overriding question is "When has a fuel supply and/or petrochemical industrial entity done enough?" This question may be particularly difficult to address when considering the wide range of agencies that may be involved in a threat situation. Other organizations, such as the Environmental Protection Agency, the Department of Energy, the Centers for Disease Control, the Department of Transportation, law enforcement agencies, public health departments, and so on, will each have unique obligations or interests in responding to a severe release or explosion threat.

When Is Enough Enough?

The guiding principle for responding to severe release or explosion threats is one of "due diligence" or "what is a suitable and sensible response to a threat?" As discussed above, some response to fuel supply and/or petrochemical contamination threats is warranted due to the public health implications of an actual contamination incident. However, an energy sector unit or petrochemical facility could spend a lot of time and money overresponding to every threat, which would be an ineffective use of resources. Furthermore, overresponse to threats carries its own adverse impacts.

Ultimately, the answer to the question of "due diligence" must be decided at the local level and will depend on a number of considerations. Among other factors, local authorities must decide what level of risk is reasonable in the context of a perceived threat. Careful planning is essential to developing an appropriate response to terrorist threats, and in fact, one primary objective of the Environmental Protection Agency's *Response Protocol Tool Box* (RPTB) is to aid users in the development of their own site-specific plans that are consistent with the needs and responsibilities of the user. Beyond planning, the RPTB considers a careful evaluation of any terrorist threat and an appropriate response based on the evaluation to be the most important element of due diligence.

In the RPTB, the threat management process is considered in three successive stages: "possible," "credible," and "confirmed." Thus, as the threat escalates through these three states, the actions that might be considered due diligence expand accordingly. The following paragraphs describe, in general terms, actions that might be considered as due diligence at these various stages.

- Stage 1: "Is the threat possible?" If a petrochemical facility is faced with a terrorism threat, it should evaluate the available information to determine whether or not the threat is "possible" (i.e., something could have actually happened). If the threat is "possible," immediate operational response actions might be implemented, and activities such as site characterization would be initiated to collect additional information to support the next stage of the threat evaluation.
- Stage 2: "Is the threat credible?" Once a threat is considered "possible," additional information will be necessary to determine whether the threat is "credible." The threshold at the credible stage is higher than that at the possible stage, and in general there must be information to corroborate the threat in order for it to be considered "credible."
- Stage 3: "Has the incident been confirmed?" Confirmation implies that definitive evidence and information have been collected to establish the presence of a threat to the fuel supply or petrochemical facility. Obviously, at this stage the concept of due diligence takes on a whole new meaning since authorities are now faced with death and destruction and a potential public health crisis. Response actions at this point include all steps necessary to protect public health, property, and the environment.

PREPARATION

As environmental safety and health consultants for various utilities and others on the East Coast, we have performed numerous pre-OSHA audit inspections and audits of various plant process safety management/risk management planning (PSM/RMP) compliance programs. During these site visits, one factor always seemed to be universal. While conducting the plant walk-around to gauge the plant's overall profile and status with OSHA and EPA compliance, we almost always find that plant managers or superintendents who accompanied us were shocked to find out what was actually going on or taking place in their facilities. They would scratch their heads and ask various workers: "What the hell are you doing? Where did that new machine come from? When was it installed? Why is that door broken? Who told you to paint that door, machine, or other apparatus? When did that hole get in the fence? Who left the back gate open? Where is the foreman?" and so on. Eventually, in an expression of utter consternation, the managers/superintendents ask, "Who the hell is in charge around here?" And this is basically the question we find ourselves asking—far too often.

In one inspection performed at a plant right after 9/11, we drove up to the entrance gate and were impressed with the height and condition of the barbed-razor-wire-topped fence and gates. We could not enter through the gates until we identified ourselves over a speaker system while a CCTV camera focused on my face. We were let in and given instructions to sign in at the main office. Not bad, just the way it should be . . . or so we thought at the time. After walking most of the plant site in the company of the plant manager, we approached the back fence area, which was close to a huge chlorine storage building. At the terminal end of the plant and fence, we noticed a large gate that was propped open with ivy growing on it, through it, and around it—obviously, the gate had been in the open position for quite some time. We asked the plant manager why the gate was open. He stated that it was always open . . . that it led to a downhill path to a beach area below where plant personnel had constructed a picnic area fronting the James River.

We walked the path to the bottom picnic area, looked around, and then looked back up the path toward the open gate and the prominent structure standing within, the chlorine storage building. While walking back up the path to the gate, we asked the plant manager whether he was not concerned about the safety of the plant because the gate was left open, and especially about the safety of the fifty tons of deadly chlorine gas stored in the chlorine storage building.

"Nah . . . no way . . . we are safe here. I really don't see anyone swimming upriver just to get into the plant site, ha. Besides, we are surrounded by woods out here . . . there's nothing to attack anyway."

Once inside the plant, we asked the plant manager whether he was not worried about terrorists or some disgruntled former employee using a boat filled with explosives or some other weapon(s) gaining easy access to the plant and especially the chlorine building via the James River beach landing and picnic area?

"Nah, that will never happen . . . who would be that stupid? There's nothing around here worth blowin' up!"

Later, when we checked the GIS system data and maps pertaining to and showing the plant and surrounding area, we noted that about one-half mile from the plant site was a large housing area, a brewery (with seven hundred–plus employees), and a very large theme and historical park—annually visited by more than 2,500,000 people each summer.

Know Your Energy Sector and Petrochemical Production Systems

All energy sector and petrochemical industry managers and process equipment operating personnel must know their plants/facilities. For these persons, there is no excuse for not knowing every square inch of the plant site. In particular, plant workers should know about any and all construction activities under way on the site; the actual construction parameters of the plant; and especially operation of all plant unit processes. In addition, plant management must not only know its operating staff but also its customers.

Construction and Operation

Each energy sector or petrochemical industrial facility is unique with respect to age, operation, and complexity. Another aspect to be considered relates to the energy-producing/using and petrochemical processing system. This is important, particularly in evaluating the potential spread of a spilled contaminant. Propagation of petrochemical contaminants through a system is dependent on a number of factors, including mixing conditions at the point of contamination, hydraulic conditions within water systems at the time of the contaminant introduction, and reactions between the contaminant and other materials in the system or environment.

Information about construction materials used in the system may be contained within the plant records and can be useful in evaluating the fate and transport of a particular petrochemical contaminant through a system. For example, a particular contaminant may adsorb to the pipe material used in a utility's distribution system, and this type of information would be critical in evaluating remediation options following a petrochemical contamination incident.

Personnel

Energy sector employees are generally its most valuable asset in preparing for and responding to petrochemical contamination threats and incidents. They have knowledge of the system and hazards associated with the petrochemicals. The importance of knowledgeable and experienced personnel is highlighted by the complexity of most petrochemical processing systems. This complexity makes a successful contamination of a specific target contingent upon detailed knowledge of the system configuration and usage patterns. If perpetrators have somehow gained a sophisticated understanding of a chemical production process, the day-to-day experience of energy production or petrochemical system personnel will prove an invaluable tool to countering any attacks. For instance, personnel may continually look for unusual aspects of daily operation that might be interpreted as a potential threat warning and may also be aware of specific characteristics of the system that make it vulnerable to petrochemical contamination or worse.

Customers

Knowledge of energy supplies and petrochemical customers is an important component of preventing and managing fires, explosions, and/or contamination incidents. Prevention is based largely on understanding potential targets of chemicals. Of special concern may be hospitals, schools, government buildings, or other institutions where large numbers of people could be directly or indirectly affected by a hazardous petrochemical threat or incident. Steps taken to protect the customer's employees and property, such as enhancements to the physical security of the customer's property at these locations, may deter the attack itself.

Energy and petrochemical customers vary significantly with regard to their expectations of what constitutes acceptable service, so it is necessary to consider the manner in which petrochemicals are used in a particular system. For example, high chemical demand that is largely driven by industry has different implications compared to high usage rates in an urban center with a high population density. Some customers, such as factories and product manufacturers, may have certain petrochemical quality requirements. Sensitive subpopulations, including children and the elderly, can exhibit adverse health effects at doses more than an order of magnitude lower than those necessary to produce serious injury or death in a healthy adult. That being said, for the purposes of managing chemical fire, explosion, and contamination threats, it is important to keep in mind that the most important goal is protecting the health of the public as a whole. Planning, preparation, and allocation of resources should be directed toward protecting the public at large, beyond specific demographic groups or individual users.

Update ERP for Intentional Spill, Explosion, or Contamination

ERPs (emergency response plans) are nothing new to energy sector facilities/plants and petrochemical industries, since many have developed ERPs to deal with natural disasters, accidents, violence in the workplace, civil unrest, and so on. Because the energy sector and its petrochemical industries are a vital ingredient of our way of life, it has been prudent for energy sector entities and petrochemical industries to develop ERPs in order to help ensure the continuous flow of water to the community. However, many energy sector and petrochemical industry ERPs developed prior to 9/11 do not explicitly deal with terrorist threats, such as intentional fire, explosion, or contamination. Recently, the U.S. Congress and federal regulators have required energy sector sites and especially petrochemical industries to prepare or revise, as necessary, an ERP to reflect the findings of their vulnerability assessment and to address terrorist threats.

Establish Communication and Notification Strategy

Communication strategies must be planned and made available to all potential participants prior to an actual incident or threat. For the purposes of responding to an energy sector or petrochemical industry threat, the communication structure could have several management levels within the industry as well as external to the industry (governmental emergency response units, for example) that may be involved in management of a threat. The hierarchy of potential participants includes the energy sector site/facility and petrochemical industry, local government, regional government (e.g., county), state government, and federal government. Not all of these levels would necessarily be involved in every situation; however, the mechanism and process through which they interact must be decided in advance of an incident to achieve optimal public health and environmental protection. For any type of petrochemical incident, CHEMTREC (CHEMical Transportation Emergency Center) is dedicated to assisting emergency responders in dealing with incidents involving a hazardous material and is available twenty-four hours per day. Due to the number and variety of possible participants, planning for effective communication is critical.

Perform Training and Desk/Field Exercises

In addition to a lack of planning, another reason that emergency response plans fail is lack of training and practice. Training provides the necessary means for everyone involved to acquire the skills to fulfill his or her role during an emergency. It may also provide important "buy-in" to the response process from both management and staff, which is essential to the success of any response plan. Desk exercises (also known as "tabletops," "sand lot," or "dry runs") along with field exercises allow participants to practice their skills. Also, these exercises will provide a test of the plan itself, revealing strengths and weaknesses that may be used to improve the overall plan. Improvements can include measures not only for intentional explosion of fuels or petrochemicals or contamination of an environment with petrochemicals, but also for other emergencies faced by the energy sector and petrochemical facility and the community at large.

Enhance Physical Security

Where possible, denying physical access to key sites within the energy sector system may act as a deterrent to a perpetrator. When we consider that many of the energy sector units such as distribution towers, lines, wind turbines, and so on are often in remote, unguarded areas, this can be a huge challenge. Criminals often seek the easiest route of attack, just like a burglar prefers a house with an open window or an open automobile with keys in the ignition. Aside from deterring actual attacks, enhancing physical security has other benefits. For example, installation of fences and locks may reduce the rate of false alarms. Without surveillance equipment or locks, it may not be possible to determine whether a suspicious individual has actually entered a vulnerable area. The presence of a lock and a determination as to whether it has been cut or broken provides sound, although not definitive, evidence that an intrusion has occurred. Likewise, security cameras can be used to review security breaches and determine whether the incident was simply due to trespassing or is a potential contamination threat. The costs of enhancing physical security may be justified by comparison to the cost of responding to just one "credible" fuel or petrochemical fire, explosion, or contamination threat involving site characterization and lab analysis for potential contaminants. Chapter 9 provides a more in-depth discussion of physical security devices.

Establish Baseline Monitoring Program

Background concentrations of suspected or tentatively identified hazardous petrochemical contaminants may be extremely important in determining whether an incident has occurred. In some cases, and for some hazardous petrochemicals, background levels may be at detectable concentrations. Baseline occurrence information is derived from monitoring data and is used to characterize typical levels of a particular petrochemical contaminant.

SITE CHARACTERIZATION AND SAMPLING

Site characterization is defined as the process of collecting information from an investigation site in order to support the evaluation of a hazardous petrochemical-based incident that could lead to a fire, explosion, or contamination threat. Site characterization promotes and recommends the scientific method of inquiry using the conceptual site modeling process to characterize the environmental condition of a site using a multiphased approach. Site characterization activities include the site evaluation, field safety screening, rapid field testing of the water, and sample collection. The investigation site is the focus of site characterization activities, and if a suspected contamination site has been identified, it will likely be designated as the primary investigation site. Additional or secondary investigation sites may be identified due to the potential spread of a suspected contaminant. The results of site characterization are of critical importance to the threat evaluation process.

There are two broad phases of site characterization: planning and implementation. The incident commander is responsible for planning, while the site characterization team is responsible for implementing the site characterization plan. This section is intended as resource for those involved in either the planning or implementation phases of site characterization. While the target audience is primarily petrochemical industry managers and staff, other organizations (e.g., police, fire departments, FBI, and EPA criminal investigators) may be involved in site characterization activities.

Site Characterization Process

The EPA's (2003) *Response Protocol Toolbox* for water contamination events is the model we use (because of its applicability) to describe the following five-stage energy sector petrochemical site characterization process:

1. *Customizing the Site Characterization Plan.* A site characterization plan is developed for a specific threat (possibly from a generic site characterization plan) and guides the team during site characterization activities.
2. *Approaching the Site.* Before entering the site, an initial assessment of site conditions and potential hazards is conducted at the site perimeter.

3. *Characterizing the Site.* The customized site characterization plan is implemented by conducting a detailed site investigation and rapid testing of the water, air, and soil.
4. *Collecting Samples.* Water/air/soil samples are collected in the event that lab analysis is required.
5. *Exiting the Site.* Following completion of site characterization, the site is secured, and personnel exit the site and undergo any necessary decontamination.

While site characterization can be considered and implemented as a discrete process, it is important to regard it as an element of the threat evaluation process. In particular, site characterization is an activity initiated in response to a "possible" accidental or intentional fire, explosion, or contamination threat in order to gather information to help determine whether or not the threat is "credible." Initially, information from the threat evaluation supports the development of the customized site characterization plan. As this plan is implemented, the observations and results from site characterization feed into the threat evaluation. In turn, the revised threat evaluation may indicate that the threat is "credible" or "not credible," or that the site characterization plan needs to be received in the field to collect more information in order to make this determination. Because threat evaluation and site characterization are interdependent, the incident commander must be in constant communication with the site characterization team while it is performing its tasks.

The first step is to develop a customized site characterization plan, which is based on the specific circumstances of the threat warning. This customized plan may be adapted from a generic site characterization plan, which is developed as part of a utility's preparation for responding to hazardous petrochemical threat. The site characterization team will use the customized plan as the basis for its activities at the investigation site. After an initial evaluation of available information, it is necessary to identify an investigation site where site characterization activities will be conducted. During the development of the customized plan, it is important to conduct an initial assessment of site hazards, which is critical to the safety of the site characterization team and may impact the makeup of the team. If there are obvious signs of hazards at the site, then teams trained in hazardous materials safety and handling techniques, such as HazMat, may need to conduct an initial hazard assessment at the site and either "clear" the site for entry by utility personnel or decide to perform all site characterization activities themselves. Obvious signs of hazards would provide a basis for determining that a threat is "credible." Furthermore, the site might be considered a crime scene if there are obvious signs of hazards, and law enforcement may take over the site investigation.

Upon arrival at the site perimeter, the team first conducts field safety screening and observes site conditions. The purpose of field safety screening activities is to identify potential environmental hazards that might pose a risk to the site characterization team. The specific field safety screening performed should be identified in the site characterization plan and might include screens for radioactivity and volatile organic chemicals (VOCs). If the team detects signs of hazard, it should stop its investigation and immediately contact the incident commander to report its findings.

If no immediate hazards are identified during approach to the site, the incident commander will likely approve the team to enter the site and perform the site characterization. During this stage, the team will continue field safety screening at the site, conduct a detailed site investigation, and perform rapid field testing of the water that is suspected of being contaminated.

Rapid field testing has three objectives: (1) provide additional information to support the threat evaluation process; (2) provide tentative identification of contaminants that would need to be confirmed later by lab testing; and (3) determine whether hazards tentatively identified in the petrochemical release require special handling precautions. The specific rapid field testing performed should be identified in the site characterization plan. Specific field testing performed should be based on the circumstances of the specific threat and should consider the training, experience, and resources of the site characterization team. Negative field test results are not a reason to forgo petrochemical sampling, since field testing is limited in scope and can result in false negatives.

Following rapid field testing, samples of the potential contaminated air/water/soil will be collected for potential lab analysis. The decision to send samples to a lab for analysis should be based on the outcome of the threat evaluation. If the threat is determined to be "credible," then samples should be immediately delivered to the lab for analysis. The analytical approach for samples collected from the site should be developed with input from the supporting lab(s), based on information from the site characterization and threat evaluation. On the other hand, if the threat is determined to be "not credible," then samples should be secured and stored for a predetermined period in the event that it becomes necessary to analyze the samples at a later time.

At this point, response actions may be implemented to protect public health. However, if the threat is determined to be "not credible," then samples may be collected, preserved, and stored in the event that it becomes necessary to analyze them later.

Upon completion of site characterization activities, the team should prepare to exit the site. At this stage, the team should make sure that it has documented its findings, collect all equipment and samples, and resecure the site (e.g., lock doors, hatches, and gates). If the site is considered to be a potentially hazardous site or crime scene, there may be additional steps involved in exiting this site.

Roles and Responsibilities

The incident commander and the site characterization team leader are key personnel in site characterization. The incident commander has overall responsibility for managing the response to the threat and is responsible for planning and directing site characterization activities. The incident commander may also approve the site characterization team to proceed with its activities at key decision points in the process (e.g., whether or not to enter the site following the approach).

The site characterization team leader is responsible for implementing the site characterization plan in the field and supervising site characterization personnel. The site characterization team leader must coordinate and communicate with the incident commander during site characterization.

Depending on the nature of the contamination threat, other agencies and organizations may be involved or otherwise assume some responsibility during planning and implementation of site characterization activities. However, it is the incident commander who has the ultimate responsibility for determining the scope of the site characterization activities and the team makeup.

Planning for Site Characterization

Providing training of staff involved in site characterization and sampling activities is critical. Responding to the site of a potential contamination incident is very different from routine inspection and sampling activities performed by utility staff. The equipment and safety procedures used at the site of a potential contamination incident may differ significantly from those used during more typical field activities. Providing staff training in the procedures presented in this section will help to ensure that they are properly and safely implemented during emergency situations.

Safety and Personnel Protection

Proper safety practices are essential for minimizing risk to the energy sector petrochemical site characterization team and must be established prior to an incident in order to be effective. Field personnel involved in site characterization activities should have appropriate safety training to conform to appropriate regulations, such as OSHA 1910.120, which deals with hazardous chemical substances. If planners and field personnel do not conclude that these regulations are applicable to them, they may still wish to adopt some of the safety principles in these regulations. The following guidance is provided to help the user develop the facility's own safety policies and practices. These safety policies should be consistent with the equipment and capabilities of the site characterization team and any applicable regulations.

The appropriate level of personal protection necessary to safely perform the site characterization activities will depend on the assessment of site hazards that might pose a risk to the site characterization team. The hazard assessment may be further refined during the approach to the site, based on the results of the field safety screening and initial observations of site conditions. Two general scenarios are considered, one in which there are no obvious signs of immediate hazards, and one in which there are indicators of site hazards.

Sample Collection Kits and Field Test Kits

Two types of kits are discussed in this section, sample collection kits and field test kits. Sample collection kits will generally contain all sample containers, materials, supplies, and forms necessary to perform sample collection activities. Field test kits contain the equipment and supplies necessary to perform field safety screening and rapid field testing of the air, water, and/or soil. Sample collection kits will generally be less expensive to construct than field test kits, and by constructing these two types of kits separately, sample collection kits can be prepositioned throughout a system while the more expensive field test kits may be assigned to specific site characterization teams or personnel.

The design and construction of sample collection and field test kits is a planning activity, since these kits must be ready to go at a moment's notice in response to a "possible" contamination threat. In addition to improving the efficacy of the site characterization and sampling activities, advance preparation of sample collection and field test kits offers several advantages:

- Sample collection and field test kits can be standardized throughout an area to facilitate sharing of kits in the event of an emergency that requires extensive sampling.
- Collection of a complete sample set is more likely to be achieved through the use of predesigned kits.
- Sample collection kits can be prepositioned at key locations to expedite the sampling process.
- Personnel responsible for site characterization can become familiar with the content of the kits and trained in the use of any specialized equipment.

Generic Site Characterization Plan

A site characterization plan is developed to provide direction and communication between the incident commander and the site characterization team, which will facilitate the safe and efficient implementation of site characterization activities. The plan should be developed expeditiously since the site characterization results are an important input to the threat evaluation process. The rapid development of a site characterization plan can be facilitated by the development of a generic site characterization plan, which is easily customized to a specific situation. While the circumstances of a particular threat warning will dictate the specifics of a customized site characterization plan, many activities and procedures will remain the same for most situations, and these common aspects can be documented in the generic site characterization plan.

Potential elements of a generic plan include pre-entry criteria, communications, team organization and responsibilities, safety, field testing, sampling, and exiting the site.

Pre-entry criteria define the conditions and circumstances under which site characterization activities will be initiated and the manner in which these activities will proceed. At each stage of the process (i.e., approach to the site, on-site characterization activities, sample collection, and exiting the site), specific criteria may be defined for proceeding to the next stage. The pre-entry criteria may also specify the general makeup of the site characterization team under various circumstances. For example, under low hazard conditions, petrochemical facility teams may perform site characterization, while specially trained responders might be called upon to assist in the case of potentially hazardous conditions at the site. The criteria developed for a particular petrochemical facility should be consistent with the role that the facility has assumed in performing site characterization activities.

The generic plan should define communication processes to ensure rapid transmittal of findings and a procedure for obtaining approval to proceed to the next stage of site characterization. It is advisable for the site characterization team to remain in constant communication with the incident commander for the entire time that it is on site. The plan should provide an approval process for the team to advance through the approach and on-site evaluation stages of the characterization, to ensure that the team is not advancing into a hazardous situation. Communication devices (e.g., cell phone, two-way radio, or panic button) can be used to alert incident command of problems/observations encountered in the field. The communication section of the generic plan should also discuss coordination with other agencies (e.g., law enforcement, fire department) and contingencies for contacting HazMat responders.

Field testing and sampling may be handled in the generic plan by presenting a menu that covers all potential options available to the utility, based on both internal and external capabilities. In developing a customized plan, the incident commander can simply check off the field tests and sampling requirements that are appropriate for the specific situation. The site characterization plan may also need to be revised in the field based on the observations of the team.

Quality Assurance for Field Testing and Sampling

Because of the diversity of potential field testing and sampling activities during the characterization, there may be no specific quality assurance (QA) activities that apply to all sampling procedures. However, the following general QA principles would apply in most cases and are consistent with the QA guidelines published by the EPA's Environmental Response Team:

- All data should be documented on field data sheets or within site logbooks.
- All instrumentation should be operated in accordance with operating instructions as supplied by the manufacturer, unless otherwise specified in the work plan. Equipment checkout and calibration activities should occur prior to site characterization and documented.
- Any relevant QA principles and plans specific to the particular water utility or responding organization should be observed.

Maintaining Crime Scene Integrity

The suspected petrochemical contamination site that is the focus of site characterization activities could potentially become the scene of a criminal investigation. If law enforcement takes responsibility for incident command because it believes a crime has been committed, it will control the site and dictate how any additional activities, such as site characterization, are performed. In cases in which the petrochemical facility is still responsible for incident command, it may still be prudent to take precautions to maintain the integrity of the potential crime scene during site characterization activities. The following guidelines for maintaining crime scene integrity are provided, although this should not necessarily be considered an exhaustive list:

- If there is substantial physical evidence of contamination at a site, the threat will likely be deemed "credible" from a utility and a law enforcement perspective. In this case, law enforcement may take control of the site and limit the activities performed by other organizations at the site.
- Substantial physical evidence of contamination might include discarded PPE, equipment (such as pumps and hoses), or containers with residual material. Special care should be taken to avoid moving or disturbing any potential physical evidence.
- Evidence should not be handled except at the direction of the appropriate law enforcement agency. Specially trained teams from the law enforcement community are best suited (and may be jurisdictionally required) for the collection of physical evidence from a contaminated crime scene.
- The collection of physical evidence is not generally considered time sensitive; however, site characterization and sampling activities are time sensitive due to the public health implications of contaminated environmental media: air, water, and/or soil. Thus, collection of environmental media samples may precede collection of physical evidence, and care must be taken not to disturb the crime scene while performing these activities. If samples can be collected outside the boundaries of the suspected crime scene, it may avoid concerns about the integrity of the crime scene.

- Water, air, and/or soil samples collected for the purpose of confirming/dismissing a contamination threat and identifying a contaminant could potentially be considered evidence and should be handled accordingly.
- Since the analytical results may be considered evidence as well, it is important to use a qualified lab for analytical support. If law enforcement has taken control of the situation prior to sample collection, it may require the collection of an additional sample set to be analyzed by its designated lab.
- Photographs and videos can be taken during the site characterization for use in the criminal investigation. Law enforcement should be consulted for proper handling during and after taking photographs/videos to ensure integrity of the evidence.

Maintaining crime scene integrity during site characterization is largely an awareness issue. If the site characterization team integrates the guidelines outlined above into its on-site activities, doing so will go a long way toward maintaining the integrity of the crime scene.

Site Characterization Protocol

This section lists procedures for conducting site characterization activities. A more in-depth treatment of this important subject area is provided by the EPA's *Response Protocol Toolbox* (2003), which is highly recommended reading. The site characterization protocol is divided into five stages:

- *Customizing the Site Characterization Plan*: Review the initial threat evaluation, review and customize the generic site characterization plan, identify the investigation site, conduct a preliminary hazard assessment, develop a sampling approach, and form the site characterization team.
- *Approaching the Site*: Establish the site zone, conduct field safety screening, and observe site conditions.
- *Characterizing the Site*: Repeat field safety screening, conduct the detailed site evaluation, and perform rapid field testing of the air, water, and/or soil.
- *Collecting Samples*: Fill sample containers, preserve samples if necessary, and initiate chain of custody.
- *Exiting the Site*: Perform final site check, remove all equipment and samples from the site, and resecure the location.

Note that documentation of the site characterization activities and findings is an ongoing effort throughout each phase and results in a site characterization report.

Customizing the Site Characterization Plan

The first stage of the site characterization process is the customization of the generic plan developed as part of planning and preparation for responding to petrochemical contamination threats. In general, the incident commander will develop the customized plan in conjunction with the site characterization team leader. The steps involved in the development of the plan include (1) performing an initial evaluation of information about the threat; (2) identifying one or more investigation sites; assessing potential site hazards; (3) developing a sampling approach; and (4) assembling a site characterization team.

Site Characterization Report

In order to provide useful information to support the threat evaluation process and the development of an analytical approach, the findings of the site characterization should be summarized in a report. This report is not intended to be a formal document, but simply a concise summary of information from the site activities that can be quickly assembled within an hour or two. The recommended content of the report includes

- General information about the site
- Information about potential site hazards
- Summary of observations from the site evaluation
- Field safety screening results, including any appropriate caveats on the results
- Rapid field water, air, and/or soil testing results, including any appropriate caveats on the results
- Inventory of samples collected and the sites from which they were collected
- Any other pertinent information developed during the site characterization

Sample Packaging and Transport

In order to perform analysis of samples beyond rapid field testing, it will be necessary to properly package the samples for transport to the appropriate labs as quickly as possible. Prompt and proper packaging and transport of samples will

- Protect the integrity of samples from changes in composition or concentration caused by bacterial growth or degradation that might occur at increased temperatures.
- Reduce the chance of leaking or breaking of sample containers that would result in loss of sample volume, loss of sample integrity, and potential exposure of personnel to hazardous substances.
- Help ensure compliance with shipping regulations.

REFERENCES AND RECOMMENDED READING

Burdick, D. L., and W. L. Leffler. 2001. *Petrochemicals in nontechnical language.* Tulsa, OK: PennWell.

CBO. 2004. *Homeland security and the private sector.* Washington, DC: Congressional Budget Office.

Deffeyes, K. S. 2001. *Hubbert's peak: The impending world oil shortage.* Princeton, NJ: Princeton University Press.

Fenichell. S. 1996. *Plastic: The making of a synthetic century.* New York: Harper Business.

Government Accountability Office. 2004. *Homeland security—chemical security.* Washington, DC: U.S. Government Accountability Office, GAO-04-482T.

Garverick, L., ed. 1994. *Corrosion in the petrochemical industry.* Materials Park, OH: ASM International.

Hargesheimer, Erika, ed. 2002. *Online monitoring for drinking water utilities cooperative research report.* Denver, CO: American Water Works Association.

Henry, K. 2002. The face of homeland security. *Government Security*, Apr. 1, 30–37.

Schierow, L. 2004. Chemical Plant Security, CRS Report for Congress, Congressional Research Service, U.S. Library of Congress, RL31530.

U.S. Environmental Protection Agency. 2001. *Protecting the nation's water supplies from terrorist attack: Frequently asked questions.* Washington, DC: United States Environmental Protection Agency.

———. 2003. *Response protocol toolbox: Planning for and responding to drinking water contamination threats and incidents.* Washington, DC: United States Environmental Protection Agency.

World Health Organization. 2004. *Public health response to biological and chemical weapons: WHO guidance.* 2nd ed. Geneva: World Health Organization.

7

SCADA

Hoover Dam as seen from Arizona side

What will you do when the power goes out?

Nobody has ever been killed by a cyberterrorist.

"Unless people are injured, there is also less drama and emotional appeal."

—Dorothy Denning

On April 23, 2000, police in Queensland, Australia stopped a car on the road and found a stolen computer and radio inside. Using commercially available technology, a disgruntled former employee had turned his vehicle into a pirate command center of sewage treatment along Australia's Sunshine Coast. The former employee's arrest solved a mystery that had troubled the Maroochy Shire wastewater system for two months. Somehow the system was leaking hundreds of thousands of gallons of putrid sewage into parks, rivers and the manicured grounds of a Hyatt Regency hotel—marine life died, the creek water turned black and the stench was unbearable for residents. Until the former employee's capture—during his 46th successful intrusion—the utility's managers did not know why.

Specialists study this case of cyber-terrorism because it is the only one known in which someone used a digital control system deliberately to cause harm. The former employee's intrusion shows how easy it is to break in—and how restrained he was with his power.

To sabotage the system, the former employee set the software on his laptop to identify itself as a pumping station, and then suppressed all alarms. The former employee was the "central control station" during his intrusions, with unlimited command of 300 SCADA nodes governing sewage and drinking water alike.

The bottom line: as serious as the former employee's intrusions were they pale in comparison with what he could have done to the fresh water system—he could have done anything he liked. (Gellman 2002)

IN THE WORDS OF MASTER SUN TZU, FROM *THE ART OF WAR*:

Those who are first on the battlefield, and await the opponents are at ease; those who are last, and head into battle are worn out.

In 2000, the FBI identified and listed threats to critical infrastructure. These threats are listed and described in table 7.1.

THE ENERGY SECTOR AND CYBERSPACE

Today's developing "information age" technology has intensified the importance of critical infrastructure protection, in which cyber security has become as critical as physical security to protecting energy sector critical infrastructure.

In the past few years, especially since 9/11, it has been somewhat routine for us to pick up a newspaper or magazine or view a television news program where a major topic of discussion is cyber security or the lack thereof. Many of the cyber intrusion incidents we read or hear about have added new terms or new uses for old terms to our vocabulary. For example, old terms such as botnets (short for robot networks, also balled bots, zombies, botnet fleets, and many others), which denote groups of computers that have been compromised with malware such as Trojan horses, worms, backdoors, remote control software, and viruses, have taken on new connotations in

Table 7.1. Threats to Critical Infrastructure Observed by the FBI

Threat	*Description*
Criminal groups	There is an increased use of cyber intrusions by criminal groups who attack systems for purposes of monetary gain.
Foreign intelligence services	Foreign intelligence services use cyber tools as part of their information gathering and espionage activities.
Hackers	Hackers sometimes crack into networks for the thrill of the challenge or for bragging rights in the hacker community. While remote hacking once required a fair amount of skill or computer knowledge, hackers can now download attack scripts and protocols from the Internet and launch them against victim sites. Thus, while attack tools have become more sophisticated, they have also become easier to use.
Hacktivists	Hacktivism refers to politically motivated attacks on publicly accessible web pages or e-mail servers. These groups and individuals overload e-mail servers and hack into websites to send a political message.
Information warfare	Several nations are aggressively working to develop information warfare doctrine, programs, and capabilities. Such capabilities enable a single entity to have a significant and serious impact by disrupting the supply, communications, and economic infrastructures that support military power—impacts that, according to the director of the Central Intelligence Agency, can affect the daily lives of Americans across the country.
Inside threat	The disgruntled organization insider is a principal source of computer crimes. Insiders may not need a great deal of knowledge about computer intrusions because their knowledge of a victim system often allows them to gain unrestricted access to cause damage to the system or to steal system data. The insider threat also includes outsourcing vendors.
Virus writers	Virus writers are posing an increasingly serious threat. Several destructive computer viruses and "worms" have harmed files and hard drives, including the Melissa Macro Virus, the Explore.Zip worm, the CIH (Chernobyl) Virus, Nimda, and Code Red.

Source: Federal Bureau of Investigation, *Threat to Critical Infrastructure* (Washington, DC: Federal Bureau of Investigation, 2000).

regard to cyber security issues. Relatively new terms such as scanners, Windows NT hacking tools, ICQ hacking tools, mail bombs, sniffers, logic bombs, nukers, dots, backdoor Trojans, key loggers, hackers' Swiss knives, password crackers, blended threats, Warhol worms, Flash threats, targeted attacks, and BIOS crackers are now commonly read or heard. New terms have evolved along with various control mechanisms. For example, because many control systems are vulnerable to attacks of varying degrees, these attack attempts range from telephone line sweeps (wardialing), to wireless network sniffing (wardriving), to physical network port scanning, and to physical monitoring and intrusion. When wireless network sniffing is performed at (or near) the target point by a pedestrian (warwalking), meaning that instead of a person being in an automotive vehicle, the potential intruder may be sniffing the network for weaknesses or vulnerabilities on foot while posing as a person walking, the person may have a handheld PDA device or laptop computer (Warwalking 2003).

Not all relatively new and universally recognizable cyber terms have sinister connotation or meaning, of course. Consider, for example, the following digital terms: backup, binary, bit, byte, CD-ROM, CPU, database, e-mail, HTML, icon, memory, cyberspace, modem, monitor, network, RAM, Wi-Fi (wireless fidelity), record, software, World Wide Web—none of these terms normally generates thoughts of terrorism in most of us.

There is, however, one digital term, SCADA, that most people have not heard of. This is not the case, however, with those who work with the nation's critical infrastructure, including energy sector infrastructure. SCADA, or Supervisory Control And Data Acquisition System (also sometimes referred to as digital control systems or process control systems), plays an important role in computer-based control systems. From coordinating music and lights in the proper sequence with spray from water fountains to controlling systems used in the drilling and refining of oil and natural gas, control systems perform many functions. Many energy distribution networks use computer-based systems to remotely control sensitive feeds, processes, and system equipment previously controlled manually. These systems (commonly known as SCADA) allow an energy sector entity (or operation) to monitor fuel tank levels, to ensure that contents are stored at correct levels, or to monitor tank levels and, as mentioned, collect data from sensors and control equipment located at remote sites. Common energy sector system sensors measure elements such as fluid level, temperature, pressure, petrochemical purity or slurry composition, and energy and petrochemical pipeline flow rates. Common energy industry system equipment includes valves, pumps, and switching devices for distribution of electricity. The critical infrastructure of many countries is increasingly dependent on SCADA systems.

WHAT IS SCADA?

If we were to ask the specialist, "What is SCADA?" the technical response could be outlined as follows:

SCADA is

- A multitier system or interfaces with multitier systems
- Used for physical measurement and control endpoints via an RTU and PLC to measure voltage, adjust a value, or flip a switch
- An intermediate processor normally based on commercial third-party operating systems—VMS, Unix, Windows, Linux
- A means of human interface, for example, with a graphical user interface (Windows GUIs)

- A communication infrastructure consisting of a variety of transport mediums such as analog, serial, Internet, radio, and Wi-Fi

How about the nonspecialist response—for the rest of us who are nonspecialists?

For the nonspecialist and the rest of us, we could simply say that SCADA is a computer-based control system that remotely controls processes previously controlled manually. The philosophy behind SCADA control systems can be summed up by the phrase "If you can measure it, you can control it." SCADA allows an operator using a central computer to supervise (control and monitor) multiple networked computers at remote locations. Each remote computer can control mechanical processes (mixers, pumps, valves, etc.) and collect data from sensors at its remote location. Thus the phrase Supervisory Control and Data Acquisition, or SCADA.

The central computer is called the master terminal unit, or MTU. The MTU has two main functions: periodically obtain data from RTUs/PLCs, and control remote devices through the operator station. The operator interfaces with the MTU using software called human-machine interface, or HMI. The remote computer is called the program logic controller (PLC) or remote terminal unit (RTU). The RTU activates a relay (or switch) that turns mechanical equipment "on" and "off." The RTU also collects data from sensors. Sensors perform measurement, and actuators perform control.

In the initial stages utilities ran wires, also known as hardwire or land lines, from the central computer (MTU) to the remote computers (RTUs). Since remote locations can be located hundreds of miles from the central location, utilities begun to use public phone lines and modems, leased telephone company lines, and radio and microwave communication. More recently, they have also begun to use satellite links, the Internet, and newly developed wireless technologies.

DID YOU KNOW?

Modern RTUs typically use a ladder-logic approach to programming due to its similarity to standard electrical circuits. An RTU that employs this ladder-logic programming is called a programmable logic controller (PLC).

Because SCADA systems' sensors provide valuable information, many utilities and other industries established "connections" between their SCADA systems and their business systems. This allowed utility/industrial management and other staff access to valuable statistics, such as chemical usage. When utilities/industries later connected their systems to the Internet, they were able to provide stakeholders/stockholders with usage statistics on the utility/industrial web pages. Figure 7.1 provides a basic illustration of a representative SCADA network. Note that firewall protection (see chapter 9) would normally be placed between Internet and business system and between business system and the MTU.

SCADA Applications in Energy Sector Systems

As stated above, SCADA systems can be designed to measure a variety of equipment operating conditions and parameters or volumes and flow rates or electricity, natural gas, and oil or oil and petrochemical mixture quality parameters, and to respond to change in those parameters either by alerting operators or by modifying

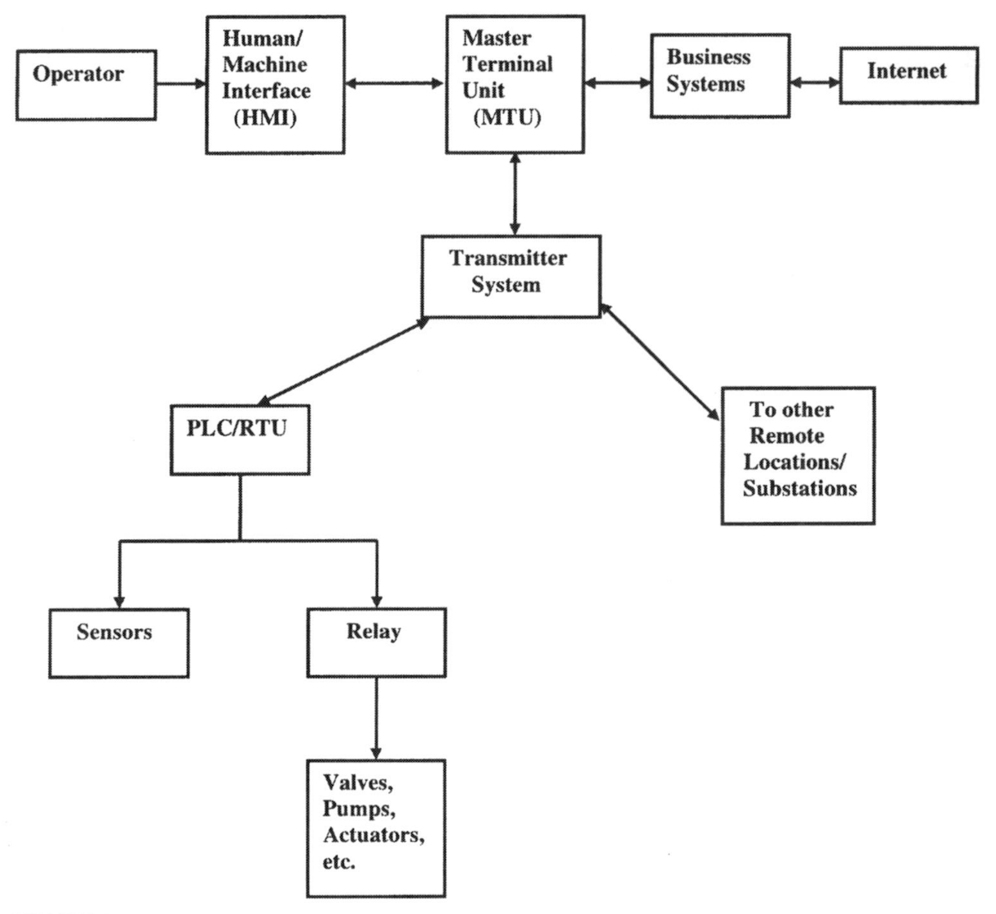

FIGURE 7.1
Representative SCADA Network

system operation through a feedback loop system without having personnel physically visit each process or piece of equipment on a daily basis to check it and/or ensure that it is functioning properly. Automation and integration of large-scale diverse assets required SCADA systems to provide the utmost in flexibility, scalability, openness, and reliability. SCADA systems are used to automate certain energy production functions; these can be performed without initiation by an operator. In addition to process equipment, SCADA systems can also integrate specific security alarms and equipment, such as cameras, motion sensors, lights, data from card-reading systems, and so on, thereby providing a clear picture of what is happening at areas throughout a facility. Finally, SCADA systems also provide constant, real-time data on processes, equipment, location access, and so on, the necessary response to be made quickly. This can be extremely useful during emergency conditions, such as when energy distribution lines or piping breaks or when potentially disruptive chemical reaction spikes appear in petrochemical processing operations.

Today, common energy sector applications for SCADA systems include, but are not limited to, those shown below.

- Bearing temperature monitor (electric generators and motors)
- Gas processing
- Electric power transmission and distribution monitoring
- Fuel oil handling system
- Hydroelectric load management
- Petroleum pilot plants
- Plant energy management
- Plant monitoring
- Power distribution monitoring
- Process controls
- Process stimulators
- Safety parameter display systems
- Tank controls
- Utility monitoring
- Turbine controls
- Turbine monitoring
- Virtual annunciator panels

Because these systems can monitor multiple processes, equipment, and infrastructure and then provide quick notification of, or response to, problems or upsets, SCADA systems typically provide the first line of detection for atypical or abnormal conditions. For example, a SCADA system connected to sensors that measure specific energy quality parameters are measured outside of a specific range. A real-time customized operator interface screen could display and control critical systems monitoring parameters.

The system could transmit warning signals back to the operators, such as by initiating a call to a personal pager. This might allow the operators to initiate actions to prevent power outages or contamination and disruption of the energy supply. Further automation of the system could ensure that the system initiated measures to rectify the problem. Preprogrammed control functions (e.g., shutting a valve, controlling flow, throwing a switch, or adding chemicals) can be triggered and operated based on SCADA utility.

SCADA VULNERABILITIES

U.S. Electric Grid Gets Hacked Into
The Associated Press (AP) reported April 9, 2009 that spies hacked into the U.S. energy grid and left behind computer programs (Trojan Horses) that would enable them to disrupt service, exposing potentially catastrophic vulnerabilities in key pieces of national infrastructure.

Even though terrorists, domestic and/or foreign, tend to aim their main focus around the critical devices that control actual energy production and delivery, according to the EPA (2005), SCADA networks were developed with little attention paid to security, making the security of these systems often weak. Studies have found that while technological advancements introduced vulnerabilities, many energy sector plans/sites and utilities have spent little time securing their SCADA networks. As a result, many SCADA networks may be susceptible to attacks and misuse. SCADA systems languished in obscurity, and this was the essence of their security, that is, until technological developments transformed SCADA from a backroom operation to a front-and-center visible control system.

Remote monitoring and supervisory control of processes began to develop in the early 1960s and adopted many technological advancements. The advent of minicomputers made it possible to automate a vast number of once manually operated switches. Advancements in radio technology reduced the communication costs associated with installing and maintaining buried cable in remote areas. SCADA systems continued to adopt new communication methods, including satellite and cellular. As the price of computers and communications dropped, it became economically feasible to distribute operations and to expand SCADA networks to include even smaller facilities.

Advances in information technology and the necessity of improved efficiency have resulted in increasingly automated and interlinked infrastructures and created new vulnerabilities due to equipment failure, human error, weather and other natural causes, and physical and cyber attacks. Some areas and examples of possible SCADA vulnerabilities include

- Human—People can be tricked or corrupted and may commit errors.
- Communications—Messages can be fabricated, intercepted, changed, deleted, or blocked.
- Hardware—Security features are not easily adapted to small self-contained units with limited power supplies.
- Physical—Intruders can break into a facility to steal or damage SCADA equipment.
- Natural—Tornadoes, floods, earthquakes, and other natural disasters can damage equipment and connections.
- Software—Programs can be poorly written.

A study included a survey that found that many water utilities were doing little to secure their SCADA network vulnerabilities (Ezell, 1998). For example, many respondents reported that they had remote access, which can allow an unauthorized person to access the system without being physically present. More than 60 percent of the respondents believed that their systems were not safe from unauthorized access and use. Twenty percent of the respondents even reported known attempts, successful unauthorized access, or use of their system. Yet twenty-two of forty-three respondents reported that they do not spend any time ensuring their network is safe, and eighteen of forty-three respondents reported that they spend less than 10 percent of their time ensuring network safety.

SCADA system computers and their connections are susceptible to different types of information system attacks and misuse, such as system penetration and unauthorized access to information. The Computer Security Institute and Federal Bureau of Investigation conduct an annual Computer Crime and Security Survey (FBI 2004). The survey reported on ten types of attacks or misuse and reported that virus and denial of service had the greatest negative economic impact. The same study also found that 15 percent of the respondents reported abuse of wireless networks, which can be a SCADA component. On average, respondents from all sectors did not believe that their organization invested enough in security awareness. Utilities as a group reported a lower average computer security expenditure/investment per employee than many other sectors such as transportation, telecommunications, and financial.

Sandia National Laboratories' *Common Vulnerabilities in Critical Infrastructure Control Systems* described some of the common problems it has identified in the following five categories (Stamp et al. 2003):

1. **System Data**—Important data attributes for security include availability, authenticity, integrity, and confidentiality. Data should be categorized according to its sensitivity, and ownership and responsibility must be assigned. However, SCADA data is often not classified at all, making it difficult to identify where security precautions are appropriate (for example, which communication links to secure, databases requiring protection, etc.).
2. **Security Administration**—Vulnerabilities emerge because many systems lack a properly structured security policy (security administration is notoriously lax in the case of control systems), equipment and system implementation guides, configuration management, training, and enforcement and compliance auditing.
3. **Architecture**—Many common practices negatively affect SCADA security. For example, while it is convenient to use SCADA capabilities for other purposes such as fire and security systems, these practices create single points of failure. Also, the connection of SCADA networks to other automation systems and business networks introduces multiple entry points for potential adversaries.
4. **Network** (including communication links)—Legacy systems' hardware and software have very limited security capabilities, and the vulnerabilities of contemporary systems (based on modern information technology) are publicized. Wireless and shared links are susceptible to eavesdropping and data manipulation.
5. **Platforms**—Many platform vulnerabilities exist, including default configurations retained, poor password practices, shared accounts, inadequate protection for hardware, and nonexistent security monitoring controls. In most cases, important security patches are not installed, often due to concern about negatively impacting system operation; in some cases technicians are contractually forbidden from updating systems by their vendor agreements.

The following incident helps to illustrate some of the risks associated with SCADA vulnerabilities.

- During the course of conducting a vulnerability assessment, a contractor stated that personnel from his company penetrated the information system of a utility within minutes. Contractor personnel drove to a remote substation and noticed a wireless network antenna. Without leaving their vehicle, they plugged in their wireless radios and connected to the network within five minutes. Within twenty minutes they had mapped the network, including SCADA equipment, and accessed the business network and data. This illustrates what a cyber security advisor from Sandia National Laboratories specializing in SCADA stated, that utilities are moving to wireless communication without understanding the added risks.

The Increasing Risk

According to the GAO (2003), historically, security concerns about control systems (SCADA included) were related primarily to protecting against physical attack and misuse of refining and processing sites or distribution and holding facilities. However, more recently there has been a growing recognition that control systems are now vulnerable to cyber attacks from numerous sources, including hostile governments, terrorist groups, disgruntled employees, and other malicious intruders.

In addition to control system vulnerabilities mentioned earlier, several factors have contributed to the escalation of risk to control systems, including (1) the adoption of standardized technologies with known vulnerabilities, (2) the connectivity of control systems to other networks, (3) constraints on the implementation of existing security technologies and practices, (4) insecure remote connections, and (5) the widespread availability of technical information about control systems.

Adoption of Technologies with Known Vulnerabilities

When a technology is not well known, not widely used, and not understood or publicized, it is difficult to penetrate it and thus disable it. Historically, proprietary hardware, software, and network protocols made it difficult to understand how control systems operated—and therefore how to hack into them. Today, however, to reduce costs and improve performance, organizations have been transitioning from proprietary systems to less expensive, standardized technologies such as Microsoft's Windows and Unix-like operating systems and the common networking protocols used by the Internet. These widely used standardized technologies have commonly known vulnerabilities, and sophisticated and effective exploitation tools are widely available and relatively easy to use. As a consequence, both the number of people with the knowledge to wage attacks and the number of systems subject to attack have increased. Also, common communication protocols and the emerging use of Extensible Markup Language (commonly referred to as XML) can make it easier for a hacker to interpret the content of communications among the components of a control system.

Control systems are often connected to other networks—enterprises often integrate their control system with their enterprise networks. This increased connectivity has significant advantages, including providing decision makers with access to real-time information and allowing engineers to monitor and control the process control system from different points on the enterprise network. In addition, the enterprise networks are often connected to the networks of strategic partners and to the Internet. Further, control systems are increasingly using wide area networks and the Internet to transmit data to their remote or local stations and individual devices. This convergence of control networks with public and enterprise networks potentially exposes the control systems to additional security vulnerabilities. Unless appropriate security controls are deployed in the enterprise network and the control system network, breaches in enterprise security can affect the operation of control systems.

According to industry experts, the use of existing security technologies, as well as strong user authentication and patch management practices, are generally not implemented in control systems because control systems operate in real time, typically are not designed with cyber security in mind, and usually have limited processing capabilities.

Existing security technologies such as authorization, authentication, encryption, intrusion detection, and filtering of network traffic and communications require more bandwidth, processing power, and memory than control system components typically have. Because controller stations are generally designed to do specific tasks, they use low-cost, resource-constrained microprocessors. In fact, some devices in the electrical industry still use the Intel 8088 processor, introduced in 1978. Consequently, it is difficult to install existing security technologies without seriously degrading the performance of the control system.

Further, complex passwords and other strong password practices are not always used to prevent unauthorized access to control systems, in part because this could hinder a rapid response to safety procedures during an emergency. As a result, according to experts, weak passwords that are easy to guess, shared, and infrequently change are reportedly common in control systems, including the use of default passwords or even no passwords at all.

In addition, although modern control systems are based on standard operating systems, they are typically customized to support control system applications. Consequently, vendor-provided software patches are generally either incompatible or cannot be implemented without compromising service, shutting down "always-on" systems, or affecting interdependent operations.

Potential vulnerabilities in control systems are exacerbated by insecure connections. Organizations often leave access links—such as dial-up modems to equipment and control information—open for remote diagnostics, maintenance, and examination of system status. Such links may not be protected with authentication of encryption, which increases the risk that hackers could us these insecure connections to break into remotely controlled systems. Also, control systems often use wireless communications systems, which are especially vulnerable to attack, or leased lines that pass through commercial telecommunications facilities. Without encryption to protect data as it flows through these insecure connections or authentication mechanisms to limit access, there is limited protection for the integrity of the information being transmitted.

Public information about infrastructures and control systems is available to potential hackers and intruders. The availability of this infrastructure and vulnerability data was demonstrated by a university graduate student, whose dissertation reportedly mapped every business and industrial sector in the American economy to the fiber-optic network that connects them—using material that was available publicly on the Internet, none of which was classified. Many of the electric utility officials who were interviewed for the National Security Telecommunications Advisory Committee's Information Assurance Task Force's Electric Power Risk Assessment expressed concern over the amount of information about their infrastructure that is readily available to the public.

In the electric power industry, open sources of information—such as product data and educational videotapes from engineering associations—can be used to understand the basics of the electrical grid. Other publicly available information—including filings of the Federal Energy Regulatory Commission (FERC), industry publications, maps, and material available on the Internet—is sufficient to allow someone to identify the most heavily loaded transmission lines and the most critical substations in the power grid.

In addition, significant information on control systems is publicly available—including design and maintenance documents, technical standards for the interconnection of control systems and RTUs, and standards for communication among control devices—all of which could assist hackers in understanding the systems and how to attack them. Moreover, there are numerous former employees, vendors, support contractors, and other end users of the same equipment worldwide with inside knowledge of the operation of control systems.

Cyber Threats to Control Systems

There is a general consensus—and increasing concern—among government officials and experts on control systems about potential cyber threats to the control systems that govern our critical infrastructures. As components of control systems increasingly make critical decisions that were once made by humans, the potential effect of a cyber threat becomes more devastating. Such cyber threats could come from numerous sources, ranging from hostile governments and terrorist groups to disgruntled employees and other malicious intruders. Based on interviews and discussions with representatives throughout the electric power industry, the Information Assurance Task Force of the National Security Telecommunications Advisory Committee concluded that an organization with sufficient resources, such as a foreign intelligence service or a well-supported terrorist group, could conduct a structured attack on the electric power grid electronically, with a high degree of anonymity and without having to set foot in the target nation.

In July 2002, the National Infrastructure Protection Center (NIPC) reported that the potential for compound cyber and physical attacks, referred to as "swarming attacks," is an emerging threat to the U.S. critical infrastructure. As NIPC reports, the effects of a swarming attack include slowing or complicating the response to a physical attack. For instance, a cyber attack that disabled the water supply or the electrical system in conjunction with a physical attack could deny emergency services the necessary resources to manage the consequences—such as controlling fires, coordinating actions, and generating light.

Control systems, such as SCADA, can be vulnerable to cyber attacks. Entities or individuals with malicious intent might take one or more of the following actions to successfully attack control systems:

- Disrupt the operation of control systems by delaying or blocking the flow of information through control networks, thereby denying availability of the networks to control system operations.
- Make unauthorized changes to programmed instructions in PLCs, RTUs, or DCS controllers, change alarm thresholds, or issue unauthorized commands to control equipment, which could potentially result in damage to equipment (if tolerances are exceeded), premature shutdown of processes (such as prematurely shutting down transmission lines), or even disabling of control equipment.
- Send false information to control system operators either to disguise unauthorized changes or to initiate inappropriate actions by system operators.
- Modify the control system software, producing unpredictable results.
- Interfere with the operation of safety systems.

In addition, in control systems that cover a wide geographic area, the remote sites are often unstaffed and may not be physically monitored. If such remote systems are physically breached, the attackers could establish a cyber connection to the control network.

Securing Control Systems

Several challenges must be addressed to effectively secure control systems against cyber threats. These challenges include (1) the limitations of current security technologies in securing control systems; (2) the perception that securing control systems may not be economically justifiable; and (3) the conflicting priorities within organizations regarding the security of control systems.

A significant challenge in effectively securing control systems is the lack of specialized security technologies for these systems. The computing resources in control systems that are needed to perform security functions tend to be quite limited, making it very difficult to use security technologies within control system networks without severely hindering performance.

Securing control systems may not be perceived as economically justifiable. Experts and industry representatives have indicated that organizations may be reluctant to spend more money to secure control systems. Hardening the security of control systems would require industries to expend more resources, including acquiring more personnel, providing training for personnel, and potentially prematurely replacing current systems that typically have a life span of about twenty years.

Finally, several experts and industry representatives indicated that the responsibility for securing control systems typically includes two separate groups: IT security personnel and control system engineers and operators. IT security personnel tend to focus on securing enterprise systems, while control system engineers and operators tend to be more concerned with the reliable performance of their control systems. Further, they indicate that, as a result, those two groups do not always fully understand each other's requirements and collaborate to implement secure control systems.

STEPS TO IMPROVE SCADA SECURITY

The President's Critical Infrastructure Protection Board and the Department of Energy (DOE) have developed the steps outlined below to help organizations improve the security of their SCADA networks. DOE (2001) points out that these steps are not meant to be prescriptive or all inclusive. However, they do address essential actions to be taken to improve the protection of SCADA networks. The steps are divided into two categories: specific actions to improve implementation, and actions to establish essential underlying management processes and policies.

Twenty-One Steps to Increase SCADA Security (Department of Energy 2001)

The following steps focus on specific actions to be taken to increase the security of SCADA networks:

1. **Identify all connections to SCADA networks.**
   Conduct a thorough risk analysis to assess the risk and necessity of each connection to the SCADA network. Develop a comprehensive understanding of all connections to the SCADA network and how well those connections are protected. Identify and evaluate the following types of connections:

   - Internal local area and wide area networks, including business networks
   - The Internet
   - Wireless network devices, including satellite uplinks

- Modem or dial-up connections
- Connections to business partners, vendors, or regulatory agencies

2. **Disconnect unnecessary connections to the SCADA network.**
To ensure the highest degree of security of SCADA systems, isolate the SCADA network from other network connections to as great a degree as possible. Any connection to another network introduces security risks, particularly if the connection creates a pathway from or to the Internet. Although direct connections with other networks may allow important information to be passed efficiently and conveniently, insecure connections are simply not worth the risk; isolation of the SCADA network must be a primary goal to provide needed protection. Strategies such as utilization of "demilitarized zones" (DMZs) and data warehousing can facilitate the secure transfer of data from the SCADA network to business networks. However, they must be designed and implemented properly to avoid introduction of additional risk through improper configuration.
3. **Evaluate and strengthen the security of any remaining connections to the SCADA network.**
Conduct penetration testing or vulnerability analysis of any remaining connections to the SCADA network to evaluate the protection posture associated with these pathways. Use this information in conjunction with risk management processes to develop a robust protection strategy for any pathways to the SCADA network. Since the SCADA network is only as secure as its weakest connecting point, it is essential to implement firewalls, intrusion detection systems (IDSs), and other appropriate security measures at each point of entry. Configure firewall rules to prohibit access from and to the SCADA network, and be as specific as possible when permitting approved connections. For example, an independent system operator (ISO) should not be granted "blanket" network access simply because there is a need for a connection to certain components of the SCADA system. Strategically place IDSs at each entry point to alert security personnel of potential breaches of network security. Organization management must understand and accept responsibility or risks associated with any connection to the SCADA network.
4. **Harden SCADA networks by removing or disabling unnecessary services.**
SCADA control servers built on commercial or open-source operating systems can be exposed to attack through default network services. To the greatest degree possible, remove or disable unused services and network daemons to reduce the risk of direct attack. This is particularly important when SCADA networks are interconnected with other networks. Do not permit a service or feature on a SCADA network unless a thorough risk assessment of the consequences of allowing the service/feature shows that the benefits of the service/feature far outweigh the potential for vulnerability exploitation. Examples of services to remove from SCADA

networks include automated meter reading/remote billing systems, e-mail services, and Internet access. An example of a feature to disable is remote maintenance. Numerous secure configurations such as the National Security Agency's series of security guides. Additionally, work closely with SCADA vendors to identify secure configurations and coordinate any and all changes to operational systems to ensure that removing or disabling services does not cause downtime, interruption of service, or loss of support.

5. **Do not rely on proprietary protocols to protect your system.**
   Some SCADA systems are unique, proprietary protocols for communications between field devices and servers. Often the security of SCADA systems is based solely on the secrecy of these protocols. Unfortunately, obscure protocols provide very little "real" security. Do not rely on proprietary protocols or factory default configuration settings to protect your system. Additionally, demand that vendors disclose any backdoors or vendor interfaces to your SCADA systems, and expect them to provide systems that are capable of being secured.
6. **Implement the security features provided by device and system vendors.**
   Older SCADA systems (most systems in use) have no security features whatsoever. SCADA system owners must insist that their system vendor implement security features in the form of product patches or upgrades. Some newer SCADA devices are shipped with basic security features, but these are usually disabled to ensure ease of installation. Analyze each SCADA device to determine whether security features are present. Additionally, factory default security settings (such as in computer network firewalls) are often set to provide maximum usability, but minimal security. Set all security features to provide the maximum security only after a thorough risk assessment of the consequences of reducing the security level.
7. **Establish strong controls over any medium that is used as a backdoor into the SCADA network.**
   Where backdoors or vendor connections do exist in SCADA systems, strong authentication must be implemented to ensure secure communications. Modems, wireless, and wired networks used for communications and maintenance represent a significant vulnerability to the SCADA network and remote sites. Successful "war dialing" or "war driving" attacks could allow an attacker to bypass all other controls and have direct access to the SCADA network or resources. To minimize the risk of such attacks, disable inbound access and replace it with some type of callback system.
8. **Implement internal and external intrusion detection systems and establish 24/7 incident monitoring.**
   To be able to effectively respond to cyber attacks, establish an intrusion detection strategy that includes alerting network administrators of malicious network activity

originating from internal or external sources. Intrusion detection system monitoring is essential 24/7; this capability can be easily set up through a pager. Additionally, incident response procedures must be in place to allow an effective response to any attack. To complement network monitoring, enable logging on all systems and audit system logs daily to detect suspicious activity as soon as possible.

9. **Perform technical audits of SCADA devices and networks and any other connected networks to identify security concerns.**
   Technical audits of SCADA devices and networks are critical to ongoing security effectiveness. Many commercial and open-sourced security tools are available that allow system administrators to conduct audits of their systems/networks to identify active services, patch level, and common vulnerabilities. The use of these tools will not solve systemic problems but will eliminate the "paths of least resistance" that an attacker could exploit. Analyze identified vulnerabilities to determine their significance, and take corrective actions as appropriate. Track corrective actions and analyze this information to identify trends. Additionally, retest systems after corrective actions have been taken to ensure that vulnerabilities were actually eliminated. Scan nonproduction environments actively to identify and address potential problems.
10. **Conduct physical security surveys and assess all remote sites connected to the SCADA network to evaluate their security.**
   Any location that has a connection to the SCADA network is a target, especially unmanned or unguarded remote sites. Conduct a physical security survey and inventory access points at each facility that has a connection to the SCADA system. Identify and assess any source of information, including remote telephone/computer network/fiber-optic cables that could be tapped; radio and microwave links that are exploitable; computer terminals that could be accessed; and wireless local area network access points. Identify and eliminate single points of failure. The security of the site must be adequate to detect or prevent unauthorized access. Do not allow "live" network access points at remote, unguarded sites simply for convenience.
11. **Establish SCADA "Red Teams" to identify and evaluate possible attack scenarios.**
   Establish a "Red Team" to identify potential attack scenarios and evaluate potential system vulnerabilities. Use a variety of people who can provide insight into weaknesses of the overall network, SCADA system, physical systems, and security controls. People who work on the system every day have great insight into the vulnerabilities of your SCADA network and should be consulted when identifying potential attack scenarios and possible consequences. Also, ensure that the risk from a malicious insider is fully evaluated, given that this represents one of

the greatest threats to an organization. Feed information resulting from the "Red Team" evaluation into risk management processes to assess the information and establish appropriate protection strategies.

The following steps focus on management actions to establish an effective cyber security program:

12. **Clearly define cyber security roles, responsibilities, and authorities for managers, system administrators, and users.**
    Organization personnel need to understand the specific expectations associated with protecting information technology resources through the definition of clear and logical roles and responsibilities. In addition, key personnel need to be given sufficient authority to carry out their assigned responsibilities. Too often, good cyber security is left up to the initiative of the individual, which usually leads to inconsistent implementations and ineffective security. Establish a cyber security organizational structure that defines roles and responsibilities and clearly identifies how cyber security issues are escalated and who is notified in an emergency.
13. **Document network architecture and identify systems that serve critical functions or contain sensitive information that require additional levels of protection.**
    Develop and document robust information security architecture as part of a process to establish an effective protection strategy. It is essential that organizations design their network with security in mind and continue to have a strong understanding of their network architecture throughout its life cycle. Of particular importance, an in-depth understanding of the functions that the systems perform and the sensitivity of the stored information is required. Without this understanding, risk cannot be properly assessed, and protection strategies may not be sufficient. Documenting the information security architecture and its components is critical to understanding the overall protection strategy and identifying single points of failure.
14. **Establish a rigorous, ongoing risk management process.**
    A thorough understanding of the risks to network computing resources from denial-of-service attacks and the vulnerability of sensitive information to compromise is essential to an effective cyber security program. Risk assessments from the technical basis of this understanding are critical to formulating effective strategies to mitigate vulnerabilities and preserve the integrity of computing resources. Initially, perform a baseline risk analysis based on current threat assessment to use for developing a network protection strategy. Due to rapidly changing technology and the emergence of new threats on a daily basis, an ongoing risk assessment process is also needed so that routine changes can be made to the protection strategy to

ensure it remains effective. Fundamental to risk management is identification of residual risk with a network protection strategy in place and acceptance of that risk by management.

15. **Establish a network protection strategy based on the principle of defense in depth.**
    A fundamental principle that must be part of any network protection strategy is defense in depth. Defense in depth must be considered early in the design phase of the development process and must be an integral consideration in all technical decision making associated with the network. Utilize technical and administrative controls to mitigate threats from identified risks to as great a degree as possible at all levels of the network. Single points of failure must be avoided, and cyber security defense must be layered to limit and contain the impact of any security incidents. Additionally, each layer must be protected against other systems at the same layer. For example, to protect against the inside threat, restrict users to access only those resources necessary to perform their job functions.
16. **Clearly identify cyber security requirements.**
    Organizations and companies need structured security programs with mandated requirements to establish expectations and allow personnel to be held accountable. Formalized policies and procedures are typically used to establish and institutionalize a cyber security program. A formal program is essential for establishing a consistent, standards-based approach to cyber security through an organization and eliminates sole dependence on individual initiative. Policies and procedures also inform employees of their specific cyber security responsibilities and the consequences of failing to meet those responsibilities. They also provide guidance regarding actions to be taken during a cyber security incident and promote efficient and effective actions during a time of crisis. As part of identifying cyber security requirements, include user agreements and notification and warning banners. Establish requirements to minimize the threat from malicious insiders, including the conducting background checks and limiting network privileges to those absolutely necessary.
17. **Establish effective configuration management processes.**
    A fundamental management process needed to maintain a secure network is configuration management. Configuration management needs to cover both hardware configurations and software configurations. Changes to hardware or software can easily introduce vulnerabilities that undermine network security. Processes are required to evaluate and control any change to ensure that the network remains secure. Configuration management begins with well-tested and documented security baselines for your various systems.
18. **Conduct routine self-assessments.**

Robust performance evaluation processes are needed to provide organizations with feedback on the effectiveness of cyber security policy and technical implementation. A sign of a mature organization is one that is able to self-identify issues, conduct root cause analyses, and implement effective corrective actions that address individual and systemic problems. Self-assessment processes that are normally part of an effective cyber security program include routine scanning for vulnerabilities, automated auditing of the network, and self-assessments of organizational and individual performance.

19. **Establish system backups and disaster recovery plans.**
Establish a disaster recovery plan that allows for rapid recovery from any emergency (including a cyber attack). System backups are an essential part of any plan and allow rapid reconstruction of the network. Routinely exercise disaster recovery plans to ensure that they work and that personnel are familiar with them. Make appropriate changes to disaster recovery plans based on lessons learned from exercises.

20. **Senior organizational leadership should establish expectations for cyber security performance and hold individuals accountable for their performance.**
Effective cyber security performance requires commitment and leadership from senior managers in the organization. It is essential that senior management establish an expectation for strong cyber security and communicate this to the subordinate managers throughout the organization. It is also essential that senior organizational leadership establish a structure for implementation of a cyber security program. This structure will promote consistent implementation and the ability to sustain a strong cyber security program. It is then important for individuals to be held accountable for their performance as it relates to cyber security. This includes managers, system administrators, technicians, and users/operators.

21. **Establish policies and conduct training to minimize the likelihood that organizational personnel will inadvertently disclose sensitive information regarding SCADA system design, operations, or security controls.**
Release data related to the SCADA network only on a strict, need-to-know basis, and only to persons explicitly authorized to receive such information. "Social engineering," the gathering of information about a computer or computer network via questions to naive users, is often the first step in a malicious attack on computer networks. The more information revealed about a computer or computer network, the more vulnerable the computer/network is. Never divulge data revealed to a SCADA network, including the names and contact information about the system operators/administrators, computer operating systems, and/or physical and logical locations of computers and network systems over telephones or to personnel unless they are explicitly authorized to receive such information. Any requests

for information by unknown persons need to be sent to a central network security location for verification and fulfillment. People can be a weak link in an otherwise secure network. Conduct training and information awareness campaigns to ensure that personnel remain diligent in guarding sensitive network information, particularly their passwords.

REFERENCES AND RECOMMENDED READING

Brown, A. S. 2008. *SCADA vs. the hackers.* American Society of Mechanical Engineers. www.memagazine.org/backissues/membersonly/dec02/features/scadavs/scadavs.html (accessed May 10, 2008).

Department of Energy. 2001. *21 steps to improve cyber security of SCADA networks.* Washington, DC: Department of Energy.

Ezell, B. C. 1998. *Risks of cyber attack to supervisory control and data acquisition.* Masters thesis, University of Virginia.

Federal Bureau of Investigation. 2000. *Threat to critical infrastructure.* Washington, DC: Federal Bureau of Investigation.

———. 2004. *Ninth annual computer crime and security survey.* Washington, DC: Computer Crime Institute and Federal Bureau of Investigation.

Gellman, B. 2002. Cyber-attacks by al Qaeda feared: Terrorists at threshold of using Internet as tool of bloodshed, experts say. *Washington Post,* June 27, A01.

Government Accountability Office. 2003. *Critical infrastructure protection: Challenges in securing control systems.* Washington, DC: U.S. Government Accountability Office.

National Infrastructure Protection Center. 2002. *National infrastructure protection center report.* Washington, DC: National Infrastructure Protection Center.

Stamp, J., et al. 2003. *Common vulnerabilities in critical infrastructure control systems.* 2nd ed. Albuquerque, NM: Sandia National Laboratories.

U.S. Environmental Protection Agency. 2005. EPA needs to determine what barriers prevent water systems from securing known SCADA vulnerabilities. In *Final Briefing Report,* ed. J. Harris. Washington, DC: U.S. Environmental Protection Agency.

Warwalking. 2003. http://warwalking.tribe.net. (accessed Accessed May 9, 2008).

Young, M. A. 2004. *SCADA systems security.* Bethesda, MD: SANS Institute.

8

Emergency Response

Power distribution lines from Grand Coulee Dam, Washington

"We're in uncharted territory."

—*Rudy Giuliani (September 11, 2001)*

"To secure ourselves against defeat lies in our own hands, but the opportunity of defeating the enemy is provided by the enemy himself."

—*Sun Tzu,* The Art of War

When Mayor Rudy Giuliani made the above statement to Police Commissioner Bernard Kerik at the World Trade Center site, September 11, 2001, to a point and to a degree, one of the first (and not to be forgotten) gross understatements of the twenty-first century had been uttered. Indeed, for citizens of the United States of America, the 9/11 events placed our level of consciousness, awareness, fear, and questions of what to do next in "uncharted territory." Actually, when you get right down to it, 9/11 generated more questions than anything else. Many are still asking the following questions today:

Why?

Why would anyone have the audacity to attack the U.S.?

What kind of cold-blooded killers would even think of conducting such an event?

Who were those Islamic radicals who perpetrated 9/11?

What were the terrorists' goal(s)?

Why?

Why were we not ready for such an attack?

Why had we not foreseen such an event?

Why were our emergency responders so undermanned, ill prepared, and ill equipped to handle such a disaster?

What took the military fighter planes so long to respond?

What did our government really know (if anything) before the events occurred?

Could anyone have prevented it?

Bottom line questions: Why us? Hell, why anyone?

Why?

These and several other questions continue to resonate today; no doubt they will continue to haunt us for some time to come.

Maybe we ask post-9/11-related questions because of who we are, what we are, and what we are not. That is, because we are Americans we are free, uninhibited thinkers who think what we say and say what we think—isn't America great! Most Americans are softhearted and sympathetic to those in need—compassion is the very nature and soul of being American. Americans are not born terrorists; they are not born into a terrorist regime; they are not raised with fear in their hearts—they are not afraid every time they leave their homes and go about their daily business. Suicide bombers and other like terrorists are those who occupy some other faraway place, definitely not America, and they are definitely not American. Right?

Notwithstanding exceptions to the rule, such as Timothy McVeigh (a so-called red-blooded American, born and raised in America) and that other idiot (whether a national or foreigner) who mailed the anthrax, terrorism was foreign to us.

Today, from a safety/security point of view, based on the events of 9/11 and the anthrax events, we should no longer be asking why. Instead, we should not waste our time, money, and energy asking why or in pointing a finger of blame at our government, military, 9/11 emergency responders, and/or the terrorists. We should stop asking why and shift our mindset to asking what if. The point is we need to stop feeling sorry for ourselves and accept the fact that there are folks out there who do not share our view of the American way of life. In chapter 6, in regard to security preparedness, we pointed to the need to ask what-if questions. Simply, what-if analysis is a proactive approach used to prevent or mitigate certain disasters, extreme events—those human or nature generated. Obviously, asking and properly answering what-if questions has little effect on preventing the actions of Mother Nature, such as earthquakes, tornadoes, hurricanes (Katrina-type events), and others. On the other hand, it is true that what-if questions, when properly posed and answered (with results), can reduce the death toll and overall damage caused by these natural disasters. We are certainly aware that these natural events are possible, probable, and likely, and their effects can be horrendous—beyond tragic. The irony is apparent, however, especially when we ask: how many of us are actually willing to move away from or out of earthquake zones, hurricane and tornado allies, and floodplains to live somewhere else?

The fact is we do not possess a crystal ball to foretell the future. What-if questions prepare us to react and respond to certain contingencies. And respond we must, because there are certain events we simply can't prevent. The best response to an event we can't prevent is summed up by the Boy Scout Motto: Be prepared!

ENERGY SECTOR CONTINGENCY PLANNING

Emergency response planning—or contingency planning—for extreme events has long been standard practice for safety professionals in energy sector and other industrial systems operations. And, again, when we use the term *energy sector*, we are including producing, refining, transporting, generating, transmitting, conserving, building, distributing, maintaining, and controlling energy systems and system components. Moreover, we include in this discussion, to a lesser degree, petrochemicals. All energy systems are considered critical infrastructure.

For many years, prudent practices have required consideration of the potential impact of severe natural events (forces of nature), including earthquakes, tornadoes, volcanoes, floods, hurricanes, and blizzards. These possibilities have been included in energy sector industry infrastructure emergency preparedness and disaster response planning. In addition, many energy sector production and distribution facilities have considered the potential consequences of man-caused disasters such as operator error and manufacturer's industrial equipment defects. Currently, energy sector managers and operators must also consider violence in the workplace. Moreover, at present, as this text has pointed out, there is a new focus of concern: the potential effects of intentional acts by domestic (homegrown) or international (foreign) terrorists.

As a result, the security paradigm has not necessarily changed, but instead has been radically adjusted—reasonable, necessary, and sensible accommodation for and mitigation of just about any emergency situation imaginable have been and continue to be made. Because we cannot foresee all future domestic or international acts of terrorism, we must be prepared to shift from the proactive to reactive mode on short notice—in some cases, on very short notice. Accordingly, we must be prepared to respond to, react to, and mitigate what we can't prevent. Unfortunately, there is more we can't prevent than we can prevent. In light of this, in this chapter we present in outline form the U.S. Department of Energy's Emergency Support Function (ESF) #12, which is the reactive procedure designed to deal with acts of terrorism (domestic or foreign). In addition, we also present and describe another reactive mitigation procedure, the template for a standard energy sector emergency response or contingency plan, dealing with natural and man-caused disasters, which could also be applied in response to acts of terrorism.

EMERGENCY SUPPORT FUNCTION #12—ENERGY SECTOR (U.S. DEPARTMENT OF ENERGY 2008)

Emergency Support Function (ESF) #12 (Energy) is intended to facilitate the restoration of damaged energy systems and components when activated by the secretary of homeland security for incidents requiring a coordinated federal response. Under Department of Energy (DOE) leadership, ESF #12 is an integral part of the larger DOE responsibility of maintaining continuous and reliable energy supplies for the United States through preventive measures and restoration and recovery actions.

ESF #12 collects, evaluates, and shares information on energy system damage and estimations on the impact of energy system outages within affected areas. Additionally ESF #12 provides information concerning the energy restoration process such as projected schedules, percent completion of restoration, and geographic information on the restoration. ESF #12 facilitates the restoration of energy systems through legal authorities and waivers. ESF #12 also provides technical expertise to the utilities, conducts field assessments, and assists government and private-sector stakeholders to overcome challenges in restoring the energy system.

ESF #12 Policies

- Addresses significant disruptions in energy supplies for any reason, whether caused by physical disruption of energy transmission and distribution systems, unexpected operational failure of such systems, or unusual economic or international political events.
- Addresses the impact that damage to an energy system in one geographic region may have on energy supplies, systems, and components in other regions relying on the same system. Consequently, energy supply and transportation problems can be intrastate, interstate, and international.
- Performs the federal coordination role for supporting the energy requirements associated with national special security events.
- Is the primary federal point of contact with the energy industry for information sharing and requests for assistance from private- and public-sector owners and operators.
- Maintains lists of energy-centric critical assets and infrastructures, and continuously monitors those resources to identify and mitigate vulnerabilities to energy facilities.
- Establishes policies and procedures regarding preparedness for attacks to U.S. energy sources and response and recover due to shortages and disruptions in the supply and delivery of electricity, oil, natural gas, coal, and other forms of energy and fuels that impact or threaten to impact large populations in the United States.

DID YOU KNOW?

Restoration of normal operations at energy facilities is the responsibility of the facility owners.

For those parts of the nation's energy infrastructure owned and/or controlled by DOE, DOE undertakes all preparedness, response, recovery, and mitigation activities.

Concept of Operations

ESF #12 provides the appropriate supplemental federal assistance and resources to enable restoration in a timely manner. Collectively, the primary and support agencies that comprise ESF #12

- Serve as the focal point within the federal government for receipt of information on actual or projected damage to energy supply and distribution systems and requirements for system design and operations, and on procedures for preparedness, restoration, recovery, and mitigation.

- Advise federal, state, tribal, and local authorities on priorities for energy restoration, assistance, and supply.
- Assist industry, state, tribal, and local authorities with requests for emergency response actions as required to meet the nation's energy demands.
- Assist federal departments and agencies by locating fuel for transportation, communications, emergency operations, and national defense.
- Provide guidance on the conservation and efficient use of energy to federal, state, tribal, and local governments and to the public.
- Provide assistance to federal, state, tribal, and local authorities utilizing Department of Homeland Security (DHS)/Federal Emergency Management Agency (FEMA)–established communications systems.

Actions

Preincident

- In cooperation with the energy sector, ESF #12 develops and implements methodologies and standards for physical, operational, and cyber security for the energy industry.
- ESF #12 conducts energy emergency exercises with the energy industry, federal partners, states, and tribal and local governments to prepare for energy and other emergencies.
- The private sector owns and operates the majority of the nation's energy infrastructure and participates along with the DOE in developing best practices for infrastructure design and operations.
- DOE assists the states in the preparation of state energy assurance plans to improve the reliability and resiliency of the nation's energy systems.
- ESF #12 works with the DHS/FEMA regions, the private sector, states, and tribes to develop procedures and products that improve situational awareness to effectively respond to a disruption of the energy sector.
- DOE monitors the energy infrastructure and shares information with federal, state, tribal, local, and industry officials.
- In anticipation of a disruption to the energy sector, DOE analyzes and models the potential impacts to the electric power, oil, natural gas, and coal infrastructures; analyzes the market impacts to the economy; and determines the effect the disruption has on other critical infrastructure and key resources.

Incident

- The private sector normally takes the lead in the rapid restoration of infrastructure-related services after an incident occurs. Appropriate entities of the private sector are integrated into ESF #12 planning and decision-making processes.

- Upon activation of ESF #12, DOE headquarters establishes the emergency management team and activates DOE disaster response procedures.
- DOE assesses the energy impacts of the incident, provides analysis of the extent and duration of energy shortfalls, and identifies requirements to repair energy systems.
- In coordination with DHS and state, tribal, and local governments, DOE prioritizes plans and actions for the restoration of energy during response and recovery operations.
- ESF #12 coordinates with other ESFs to provide timely and accurate energy information, recommends options to mitigate impacts, and coordinates repair and restoration of energy systems.
- ESF #12 facilitates the restoration of energy systems through legal authorities and waivers.
- DOE provides subject-matter experts to the private sector to assist in the restoration efforts. This support includes assessments of energy systems, latest technological developments in advanced energy systems, and best practices from past disruptions.
- ESF #12 coordinates preliminary damage assessments in the energy sector to determine the extent of the damage to the infrastructure and the effects of the damage on the regional and national energy system.
- ESF #12 serves as the primary source for reporting of critical infrastructure damage and operating status for the energy systems within the impacted area. The infrastructure liaison, if assigned, proactively coordinates with ESF #12 on matters relating to security, protection, and/or restoration that involve sector-specific, cross-sector, or cascading effects impacting ESF #12.

Postincident

- ESF #12 participates in postincident hazard mitigation studies to reduce the adverse effects of future disasters.
- ESF #12 assists DHS/FEMA in determining the validity of disaster-related expenses for which the energy industry is requiring reimbursement based upon the Stafford Act.
- DOE leads and participates in various best practices and lessons learned forums to ensure future disruptions are addressed in the most efficient manner possible.
- In coordination with the Pipeline and Hazardous Materials Safety Administration, ESF #12 ensures the safety and reliability of the nation's natural gas and hazardous material pipelines.

EMERGENCY RESPONSE PLANNING (ERP): STANDARD TEMPLATE

Note: The following criteria have not been established as anything other than guidelines and are offered not as definitive or official regulations or procedures but rather as informed advice (based on more than thirty years of safety, industrial hygiene, emergency contingency planning, and security experience) insofar as to the subject matter specific to both public and private sectors.

The goals of an emergency response plan (ERP) are to document and understand the steps needed to

- Rapidly restore energy producing, refining, transporting, generating, transmitting, conserving, building, distributing, maintaining, and controlling systems and components after an emergency.
- Minimize energy sector production process equipment damage.
- Minimize impact and loss to customers.
- Minimize negative impacts on public health and employee safety.
- Minimize adverse effects on the environment.
- Provide emergency public information concerning customer service.
- Provide hazardous petrochemical information for first responders and other outside agencies.

Although we are concerned with the energy sector in this text, the U.S. EPA-developed *Large Water System Emergency Response Outline: Guidance to Assist Community Water Systems in Complying with the Public Health and Bioterrorism Preparedness and Response Act of 2002* (dated July 2003), with minor adjustments, can be applied as a template for the energy sector. This template provides guidance and recommendations to aid facilities in the preparation of emergency response plans under the PL 107-188. The template is provided below.

Energy Sector ERP Template

I. Introduction

Safe and reliable operation is vital to every industrial operation. Emergency response planning (ERP) is an essential part of managing an energy sector plant, process, or entity. The introduction should identify the requirement to have a documented emergency response plan (EFP), the goal(s) of the plan (e.g., be able to quickly identify an emergency and initiate timely and effective response action, be able to quickly respond and repair damages to minimize system downtime), and how access to the plan is limited. Plans should be numbered for control. Recipients should sign and date a statement that includes (1) their ERP number, (2) an agreement not to reproduce the ERP, and (3) an affirmation that they have read the ERP.

ERPs do not necessarily need to be one document. They may consist of an overview document, individual emergency action procedures, checklists, additions to existing operations manuals, appendices, and so on. There may be separate, more detailed plans for specific incidents. There may be plans that do not include particularly sensitive information and those that do. Existing applicable documents should be referenced in the ERP (e.g., chemical risk management program, contamination response).

II. Emergency Planning Process

A. Planning partnerships

The planning process should include those parties who will need to help the energy sector in an emergency situation (e.g., first responders, law enforcement, public health officials, nearby utilities, local emergency planning committees, testing labs, etc.). Partnerships should track from the energy sector operation up through local, state, regional, and federal agencies, as applicable and appropriate, and could also document compliance with governmental requirements.

B. General emergency response policies, procedures, actions, documents

A short synopsis of the overall emergency management structure, how other industrial emergency response, contingency, and risk management plans fit into the ERP for energy sector emergencies, and applicable policies, procedures, action plans, and reference documents that should be cited. Policies should include interconnect agreements with adjacent communities and just how the ERP may affect them.

C. Scenarios

Use your vulnerability assessment (VA) findings to identify specific emergency action steps required for response, recovery, and remediation for applicable incident types. In section V of this plan, specific emergency action procedures addressing each of the incident types should be addressed.

III. Emergency Response Plan—Policies

A. System-specific information

In an emergency, energy sector industries need to have basic information for system personnel and external parties such as law enforcement, emergency responders, repair contractors/vendors, the media, and others. The information needs to be clearly formatted and readily accessible so that system staff can find and distribute it quickly to those who may be involved in responding to the emergency. Basic information that may be presented in the emergency response plan are the system's ID number, system name, system address or location, directions to the system, population served, number of service connections, system owner, and information about the person in charge of managing the emergency. Distribution maps, detailed plant drawings, site plans, source/storage/production energy locations, and operations manuals may be attached to this plan as appendices or referenced.

1. PWS ID, owner, contact person
2. Population served and service connections
3. System components
   (a) Pipes and constructed conveyances
   (b) Physical barriers
   (c) Isolation valves
   (d) Petrochemical treatment, storage, and distribution facilities
   (e) Electronic, computer, or other automated systems that are utilized by the energy sector industry
   (f) Emergency power generators (on-site and portable)
   (g) The use, storage, or handling of various petrochemicals
   (h) The operation and maintenance of such system components

B. Chain-of-command chart developed in coordination with local emergency planning committee (internal and/or external emergency responders, or both)
   1. Contact name
   2. Organization and emergency response responsibility
   3. Telephone number(s) (hardwire, cell phones, faxes, e-mail)
   4. State 24-hour emergency communications center telephone number

C. Communications procedures: who, what, when

   During most emergencies, it will be necessary to quickly notify a variety of parties, both internal and external, to the energy sector entity. Using the chain-of-command chart and all appropriate personnel from the lists below, indicate who activates the plan, the order in which notification occurs, and the members of the emergency response team. All contact information should be available for routine updating and readily available. The following lists are not intended to be all inclusive—they should be adapted to your specific needs.

   1. Internal notification lists
      (a) Operations dispatch
      (b) Energy sector plant manager
      (c) Energy sector processing manager
      (d) Energy production/storage/distribution manager
      (e) Facility managers
      (f) Chief energy engineer
      (g) Director of engineering
      (h) Data (IT) manager
      (i) Maintenance manager
      (j) Other

2. Local notification lists
   (a) Head of local government (i.e., mayor, city manager, chairman of board, etc.)
   (b) Public safety officials—fire, local law enforcement (LLE), police, EMS. If a malevolent act is suspected, LLE should be immediately notified and in turn will notify the FBI if required. The FBI is the primary agency for investigating sabotage to water systems or terrorist incidents.
   (c) Other government entities: health, schools, parks, finance, electric, and so on
3. External notification lists
   (a) State department of environmental quality (DEQ)
   (b) USEPA/USDOE/DHS
   (c) State police
   (d) State health department (lab)
   (e) Critical customers (special considerations for hospitals, federal, state, and county government centers, etc.).
   (f) Service aid
   (g) Mutual aid
   (h) Chemical information sharing and analysis center (ISAC)/CHEMTREC
   (i) Commercial customers not previously notified
4. Public/media notification: when and how to communicate
   Effective communication is a key element of emergency response, and a media or communications plan is essential to good communications. Be prepared by organizing basic facts about the crisis and your energy system. Develop key messages to use with the media that are clear, brief, and accurate. Make sure your messages are carefully planned and have been coordinated with local and state officials. Considerations should be given to establishing protocols for both field and office staff to respectfully defer questions to the utility spokesperson.

   Be prepared to list geographic boundaries of the affected area (e.g., west of highway A, east of highway B, north of highway C, and south of highway D) to ensure the public clearly understands the system boundaries.

E. Personnel safety

This should provide direction as to how operations staff, emergency responders, and the public should respond to a potential toxic release (e.g., electrical current released from downed transformers and lines), including facility evacuation, personnel accountability, proper personnel protective equipment as dictated by the risk management program and process safety management plan, and whether the nearby public should be "in-place sheltered" or evacuated.

F. Equipment

The ERP should identify equipment that can obviate or significantly lessen the impact of terrorist attacks or other intentional actions on the public health and protect the safety and supply of communities and individuals. The energy sector facility should maintain an updated inventory of current equipment and repair parts for normal maintenance work.

Because of the potential for extensive or catastrophic damage that could result from a malevolent act, additional equipment sources should be identified for the acquisition and installation of equipment and repair parts in excess of normal usage. This should be based on the results of the specific scenarios and critical assets identified in the vulnerability assessment that could be destroyed. For example, numerous pumps, vats, and mixers, specifically designed for the energy sector entities, could potentially be destroyed. A certain number of "long-lead" procurement equipment should be inventoried and the vendor information for such unique and critical equipment maintained. In addition, mutual aid agreements with other industries, and the equipment available under the agreement, should be addressed. Inventories of current equipment, repair parts, and associated vendors should be indicated under item 29, "Equipment Needs/Maintenance of Equipment" of Section IV, "Emergency Action Procedures."

G. Property protection

A determination should be made as to what energy sector producing, processing, and/or distribution operation/facility should be immediately "locked down," specific access control procedures implemented, initial security perimeter established, and a possible secondary malevolent event considered. The initial act may be a diversionary act.

H. Training, exercises, and drills

Emergency response training is essential. The purpose of the training program is to inform employees or what is expected of them during an emergency situation. The level of training on an ERP directly affects how well an energy sector facility's employees can respond to an emergency. This may take the form of orientation scenarios, tabletop workshops, functional exercises, and so on.

I. Assessment

To evaluate the overall ERP's effectiveness and to ensure that procedures and practices developed under the ERP are adequate and are being implemented, the energy sector industry staff should audit the program on a periodic basis.

IV. Emergency Action Procedures (EAPs)

These are detailed procedures used in the event of an operation emergency or malevolent act. EAPs may be applicable across many different emergencies and are typically common core elements of the overall municipality ERP (e.g., responsibilities, notifications lists, security procedures, etc.) and can be referenced.

A. Event classification/severity of emergency
B. Responsibilities of emergency director
C. Responsibilities of incident commander
D. Emergency operations center (EOC) activation
E. Division internal communications and reporting
F. External communications and notifications
G. Emergency telephone list (division internal contacts)
H. Emergency telephone list (off-site responders, agencies, state twenty-four-hour emergency phone number, and others to be notified)
I. Mutual aid agreements
J. Contact list of available emergency contractor services/equipment
K. Emergency equipment list (including inventory for each facility)
L. Security and access control during emergencies
M. Facility evacuation and lockdown and personnel accountability
N. Treatment and transport of injured personnel (including electrocution and petrochemical exposure)
O. Petrochemical records—to compare against historical results for baseline
P. List of available labs for emergency use
Q. Emergency sampling and analysis (petrochemical)
R. Water use restrictions during emergencies
S. Alternate temporary petrochemical supplies during emergencies
T. Isolation plans for petrochemical supply, treatment, storage, and distribution systems
U. Mitigation plans for neutralizing, flushing, and collecting spilled petrochemicals
V. Protection of vital records during emergencies
W. Recordkeeping and reporting (FEMA, DHS, DOT, OSHA, EPA, and other requirements) (It is important to maintain accurate financial records of

expenses associated with the emergency event for possible federal reimbursement.)

X. Emergency program training, drills, and tabletop exercises

Y. Assessment of emergency management plan and procedures

Z. Crime scene preservation training and plans

AA. Communication plans:

1. Police
2. Fire
3. Local government
4. Media
5. Others

BB. Administration and logistics, including EOC, when established

CC. Equipment needs/maintenance of equipment

DD. Recovery and restoration of operations

EE. Emergency event closeout and recovery

V. Incident-Specific Emergency Action Procedures (EAPs)

Incident-specific EAPs are action procedures that identify specific steps in responding to an operational emergency or malevolent act.

A. General response to terrorist threats (other than bomb threat and incident-specific threats)

B. Incident-specific response to man-made or technological emergencies

1. Contamination event (articulated threat with unspecified materials)
2. Contamination threat at a major event
3. Notification from health officials of potential contamination
4. Intrusion through supervisory control and data acquisition (SCADA)

C. Significant structural damage resulting from intentional act

D. Customer complaints

E. Severe weather response (snow, ice, temperature, lightning)

F. Flood response

G. Hurricane and/or tornado response

H. Fire response

I. Explosion response

J. Major vehicle accident response

K. Electrical power outage response

L. Water supply interruption response

M. Transportation accident response—barge, plane, train, semi-trailer/tanker

N. Contaminated/tampered-with water treatment chemicals

O. Earthquake response

P. Disgruntled employees response (i.e., workplace violence)

Q. Vandals response
R. Bomb threat response
S. Civil disturbance/riot/strike
T. Armed intruder response
U. Suspicious mail handling and reporting
V. Hazardous petrochemical spill release response (including material safety data sheets)
W. Cyber-security/supervisory control and data acquisition (SCADA) system attack response (other than incident specific, e.g., hacker)

VI. Next Steps
A. Plan review and approval
B. Practice and plan to update (as necessary, once every year recommended)
1. Training requirements
2. Who is responsible for conducting training, exercises, and emergency drills
3. Update and assessment requirements
4. Incident-specific requirements

VII. Annexes
A. Facility and location information
1. Facility maps
2. Facility drawings
3. Facility descriptions/layout
4. Other

VIII. References and Links
A. Department of Homeland Security—www.dhs/gov/dhspublic
B. Environmental Protection Agency—www.epa.gov
C. The American Water Works Association (AWWA)—www.awwa.org
D. The Center for Disease Control and Prevention—www.bt.cdc.gov
E. Federal Emergency management Agency—www.fema.gov
F. Local Emergency Planning Committees—www.epa.gov/ceppo/epclist.htm

OSHA AND EMERGENCY RESPONSE

Even though no single OSHA standard is dedicated specifically to the issue of planning for energy sector facility/site emergencies, all OSHA standards are written for the purpose of promoting a safe, healthy, accident-free, and hence emergency-free workplace. Therefore, OSHA standards do play a role in emergency prevention.

OSHA's standards, therefore, should be considered when developing emergency plans. A first step when developing emergency response plans is to review the guidelines presented earlier and to apply any and all applicable OSHA standards to your emergency response plan. This review of applicable guidelines and pertinent OSHA standards can help the managers of energy sector industrial facilities identify and then correct conditions that might exacerbate emergency situations before they occur.

Conception of an Emergency Response Plan

Typically, when we think of emergency response plans for the workplace, we often conjure up thoughts about the obvious. For example, the first workplace emergency that might come to mind is fire—a major concern because fire in the workplace is something that can happen, that happens more often than we might think, and because fire can be particularly devastating—in ways we know all too well. Most employees do not need to be informed about the dangers of fire. However, employers have the responsibility to do just this—to inform and train employees on fire, fire prevention, and fire protection. Many local codes go beyond this information requirement, insisting that employers develop and implement a fire emergency response and/or evacuation plan. The primary emphasis has been on the latter—evacuation. However, if the employer equips a workplace with fire extinguishers and other firefighting equipment and expects its employees to respond aggressively to extinguish workplace fires, then not only must the facility have an emergency response plan, the employer must also ensure that all company personnel called upon to fight the fire are completely trained on how to do so safely; this is stipulated in OSHA's 29 CFR 1910.156(c).157(g).

Another commonly considered workplace emergency response plan or scenario is designed and implemented for medical emergencies. Many chemical facilities satisfy this requirement simply by directing their employees to call 911 or some other emergency number whenever a medical emergency occurs in the workplace. Other facilities, though, may require employees to provide emergency first aid. When the employer chooses the employee-supplied first aid option, certain requirements must be met before any employee can legally administer first aid. First, the first-aid responder must be trained and certified to administer first aid. This training aspect must also include training on OSHA's Bloodborne Pathogen Standard. This standard requires that the employee be trained on the dangers inherent with handling and being exposed to human body fluids. The employee must also be trained on how to protect himself or herself from body fluid contamination. If the first-aid responder or anyone else is exposed to and contaminated by body fluids, the employer must make available the hepatitis B vaccine and vaccination series to all employees who have occupational exposure and postexposure evaluation and follow-up to all employees who have had an exposure incident; this requirement is stipulated in OSHA's 29 CFR 1910.1030.

A third type of emergency response plan required for implementation in selected (covered) facilities is OSHA's 29 CFR 1910.120 (Hazardous Waste Operations and

Emergency Response—HAZWOPER)—for releases of hazardous petrochemical materials. Unless the facility operator can demonstrate that the operation does not involve employee exposure or the reasonable possibility for employee exposure to safety or health hazards, the following operations are covered:

1. Clean-up operations required by a governmental body involving hazardous substances conducted at uncontrolled hazardous waste sites, state priority site lists, sites recommended by the EPA, NPL, and initial investigations of government identified sites that are conducted before the presence or absence of hazardous substance has been ascertained.
2. Corrective actions involving clean-up operations at sites covered by the Resource Conservation and Recovery Act of 1976 (RCRA).
3. Voluntary clean-up operations at sites recognized by federal, state, local, or other governmental bodies as uncontrolled hazardous waste sites.
4. Operations involving hazardous petrochemical waste conducted at treatment, storage, and disposal (TSD) facilities regulated by RCRA.
5. Emergency response operations for releases of, or substantial threats of releases of, hazardous petrochemical substances without regard to the location of the hazard.

The final requirement impacts the largest number of facilities that meet the criteria requiring full compliance with 29 CFR 1910.120 HAZWOPER, because many such facilities do not normally handle, store, treat, or dispose of hazardous petrochemical waste but do use or produce hazardous materials in their processes.

Because the use of hazardous petrochemical materials could lead to an emergency from the release or spill of such materials, facilities using these materials must develop and employ an effective site emergency response plan.

Before we discuss the basic goals of an effective emergency response plan from an OSHA compliance point of view, we should define "emergency response." Considering that individual facilities are different, with different dangers and different needs, defining emergency response is not always easy. However, for our purposes, we use the definition provided by CoVan (1995):

> Emergency response is defined as a limited response to abnormal conditions expected to result in unacceptable risk requiring rapid corrective action to prevent harm to personnel, property, or system function. (54)

CoVan makes another important point about emergency response, one critical for the site manager and/or safety professional. He points out that "although emergency response and engineering tends toward prevention, emergency response is a skill area

DID YOU KNOW?

OSHA's 29 CFR 1910.151 (a), (b), and (c), Medical Services and First Aid Standard, requires employers to (1) ensure the ready availability of medical personnel for advice and consultation on matters of health; (2) in the absence of an infirmary, clinic, or hospital in near proximity to the workplace that is used for the treatment of all injured employees, a person or persons shall be adequately trained to render first aid. Adequate first-aid supplies shall be readily available; and (3) where the eyes or body of any person may be exposed to injurious corrosive materials, suitable facilities for quick drenching of or flushing of the eyes and body shall be provided within the work area for immediate emergency use.

In auditing various industrial workplaces, we found that this particular OSHA standard is one of the most misunderstood by employers. Based on our experience, we found that OSHA's use of the words "near proximity to" and "adequate" contribute to the vagueness and ambiguity of this standard. Most employers understand the need to provide first-aid supplies in the workplace but generally think that this is the extent of their responsibilities in this regard. Moreover, many workplaces do train their personnel on basic first aid and CPR but do not require (in written job descriptions) that the trained employees respond to a workplace medical emergency.

We found that employers are reluctant to have employees respond to workplace medical emergencies because most employees do not want to respond. They do not want to be exposed to gore and the victim's pain. They also do not want to assume liability for trying to aid an injured victim.

The common response we received from workers can be summed up as follows: "When it comes to liability matters in America, let me sue you before you sue me."

that safety engineers must be familiar with both because of regulations and good engineering practice" (p. 54). "Good engineering practice"—the law by which all competent managers and safety professionals work and live.

Now that we have defined emergency response, let's move on to the basic goals of an effective emergency response plan. Most of the currently available literature on this topic generally lists the goals as twofold:

1. Minimize injury to facility personnel.
2. Minimize damage to facility and then return to normal operation as soon as possible.

Obviously, these goals make a great deal of good sense. However, you may be wondering about language used—a couple of key words: "facility personnel" and "damage to the facility." Remember that we are talking about OSHA requirements here. Under OSHA the primary emphasis is protecting the worker—protecting the worker's health and safety is OSHA's only focus.

What about people who live off site: the site's neighbors?

What about the environment?

These questions make the point we emphasize here. Again, OSHA is not normally concerned about the environment unless contamination of the environment (at the work site) might adversely impact the worker's safety and health. The neighbors? Again, OSHA's focus is the worker. One OSHA compliance officer explained to us that if the employer takes every necessary step to protect its employees from harm involving the use or production of hazardous petrochemical materials, then the surrounding community should have little to fear.

This statement is puzzling to us. We asked the same OSHA compliance officer about those incidents beyond the control of the employer—about accidents that could not only put employees in harm's way but also endanger the surrounding community. The answer? "Well, that's the EPA's bag—we only worry about the worksite and the worker."

Fortunately, OSHA, in combination with the U.S. Environmental Protection Agency has taken steps to overcome this blatant shortcoming (we like to think of it as an oversight). Under OSHA's Process Safety Management (PSM) and EPA's Risk Management Planning (RMP) directives, chemical spills and other chemical accidents that could impact both the environment and the "neighbors" have now been properly addressed. What PSM and RMP really accomplish is changing the typical twofold goal of an effective emergency response plan to a threefold goal.

Let's point out that the accomplishment of these two- or threefold goals or objectives is essential in emergency response. Accomplishing these goals or objectives requires an extensive planning effort prior to the emergency ("prior" being the keyword, because the attempts to develop an emergency response plan when a disaster is occurring or after one has occurred is both futile and stupid). The site manager and/or safety professional must never forget that while hazards in any facility can be reduced, risk is an element of everyday existence and therefore cannot be totally eliminated. The manager/safety professional's goal must be to keep risk to an absolute minimum. Again, to accomplish this, planning is critical—as well as essential—and should be accomplished well in advance of emergency, because program testing should be part of the overall program design.

We pointed out earlier that most emergency plans address fire, medical emergencies, and the accidental release or spills of petrochemical materials. Note that the development of emergency response plans should also factor in other possible emergencies—natural disasters, floods, explosions, and/or weather-related events that could occur and certainly will occur. Now emergency response to terrorist activity or threats must also be added to the list.

Site emergency response plans should include

Assessment of risk

Chain of command for dealing with emergencies

Assessment of resources

Training

Incident command procedures

Site security

Public relations

The Federal Emergency Management Agency (better known as FEMA), the U.S. Army Corps of Engineers, and several other agencies, as well as numerous publications, provide guidance on how to develop a site emergency response plan. Local agencies (such as fire departments, emergency planning commissions/agencies, HazMat teams, and local emergency planning committees [LEPCs]) also provide information on how to design a site plan. All of these agencies typically recommend that a site's plan contain the elements listed in table 8.1.

In the site manager/safety professional's effort to incorporate and manage a facility emergency response plan, and in the response itself, two elements mentioned earlier (security considerations and public relations) must be given special attention. If not handled correctly, the lack of effective security measures and/or improper public relations can turn an already disastrous incident into a megadisaster.

In planning security considerations, provision should be made to have a well-trained security team limit site access to only those people and that equipment that will assist in coping with and resolving the emergency.

Table 8.1. Site Emergency Response Plan

Emergency Response Notification	List of who to call and information to pass on when an emergency occurs
Record of Changes	Table of changes and dates for them
Table of Contents/ Introduction	The purpose, objective, scope, applications, policies, and assumptions for the plan
Emergency Response Operations	Details what actions must take place
Emergency Assistance Telephone Numbers	A current list of people and agencies who may be needed in an emergency
Legal Authority and Responsibility	References the laws and regulations that provide the authority for the plan
Chain of Command	Response organization structure and responsibilities
Disaster Assistance and Coordination	Where additional assistance may be obtained when the regular response organizations are overburdened
Procedures for Changing or Updating the Plan	Details who makes changes and how they are made and implemented
Plan Distribution	List of organizations and individuals who have been given a copy of the plan
Spill Cleanup Techniques	Detailed information about how response teams should handle cleanups
Cleanup/Disposal Resources	List of what is available, where it is obtained, and how much is available
Consultant Resources	List of special facilities and personnel who may be valuable in a response
Technical Library/ References	List of libraries and other information sources that may be valuable for those preparing, updating, or implementing the plan
Hazards Analysis	Details the kinds of emergencies that may be encountered, where they are likely to occur, what areas of the community may be affected, and the probability of occurrence
Documentation of Spill	The various incident and investigative reports on spills that have occurred
Hazardous Materials Information	Listing of hazardous materials, their properties, response data, and related information
Dry Runs	Training exercises for testing the adequacy of the plan, training personnel, and introducing changes

Source: FEMA, *Planning Guide and Checklist for Hazardous Materials Contingency Plans, FEMA-10* (Washington, DC: Federal Emergency Management Agency, 1981). Adaptation from R. L. Brauer, *Safety and Health for Engineers* (New York: Van Nostrand Reinhold, 1994).

Public relations (PR) can be a tricky enterprise. The person identified to interface with the media must have thorough knowledge of the site, process, and personnel involved. The PR person must also have access to the highest levels of site management. Otherwise, he or she will not be able to deal with the public/media effectively.

To a certain extent, the person who is ultimately responsible for giving the "go-ahead" to starting up in response to a perceived emergency is in a "damned if you do, damned if you don't" position. When Hurricane Isabel hit the East Coast in 2003 with hundred-mile-per-hour-plus winds and torrential rains, and as the effects of the storm were felt hundreds of miles away from the actual storm center, we all recognized that only slight changes in conditions could have pushed Isabel out to sea—and all that reaction to the approaching emergency would have looked like overkill. But that's both the beauty and the frustration of emergency planning. By calling in the process early, you may run the risk of looking overly cautious, or, heaven forbid, wimpy. But if the storm does hit, no one is going to remember that you made the right call—you just did your job. In general, energy sector site managers are people who would rather err on the side of caution (hopefully) . . . in other words, safety. Once the storm hits, it's too late to do much good if you weren't already geared up to go.

THE BOTTOM LINE

Because industrial emergencies (in less than extreme conditions) can seriously affect the surrounding community and environment, and because poor planning and/or panic can only make a bad situation worse and can also lead to additional injury and death, your role as energy sector site manager or site safety professional in emergency response is doubly important. A crisis out of hand can easily devastate a community—and your organization is (or should be) an active member of your community. By ensuring less-than-effective emergency response, energy site managers endanger not only themselves and their organizations but endanger their organization's community and standing as well.

REFERENCES AND RECOMMENDED READING

Brauer, R. L. *Safety and health for engineers.* New York: Van Nostrand Reinhold, 1994.

CoVan, J. 1995. *Safety engineering.* New York: Wiley.

Federal Emergency Management Agency. 1981. *Planning guide and checklist for hazardous materials contingency plans.* FEMA-10. Washington, DC: Federal Emergency Management Agency.

Healy, R. J. 1969. *Emergency and disaster planning.* New York: Wiley.

Office of the Federal Register. 1987. 29 CFR 1910.120. Washington, DC: Office of the Federal Register.

Smith, A. J. 1980. *Managing hazardous substances accidents.* New York: McGraw-Hill.

Spellman, F. R. 1997. *A guide to compliance for process safety management planning (PSM/RMP).* Lancaster, PA: Technomic.

U.S. Army Corps of Engineers. 1987. *Safety and health requirements manual.* Rev. ed. EM 385-1-1. Washington, DC: U.S. Army Corps of Engineers.

U.S. Department of Energy. 2008. *Emergency support function #12—Energy annex.* Washington, DC: U.S. Department of Energy.

U.S. Environmental Protection Agency. 2002. *Water utility response, recovery & remediation guidance for man-made and/or technological emergencies.* Washington, DC: U.S. Environmental Protection Agency.

———. 2003. *Large water system emergency response plan outline: Guidance to assist community water systems in complying with the Public Health Security and Bioterrorism Preparedness and Response Act of 2002.* EPA 810-F-03-007. www.epa.gov/safewater/security (accessed June 2006).

9

Security Techniques and Hardware

Coal-powered generating plant, Arizona

"There are but two powers in the world, the sword and the mind. In the long run the sword is always beaten by the mind."

—Napoleon Bonaparte

Ideally, in a perfect world, all energy sector sites/facilities would be secured in a layered fashion (a.k.a. the barrier approach). Layered security systems are vital. Using the protection "in depth" principle, requiring that an adversary defeat several protective barriers or security layers to accomplish its goal, energy sector industry infrastructure can be made more secure. Protection in depth is a term commonly used by the military to describe security measures that reinforce one another, masking the defense mechanisms from the view of intruders and allowing the defender time to respond to intrusion or attack.

A prime example of the use of the multibarrier approach to ensure security and safety is demonstrated by the practices of the bottled water industry. In the aftermath of 9/11 and the increased emphasis on homeland security, a shifted paradigm of national security and vulnerability awareness has emerged. Recall that in the immediate aftermath of the 9/11 tragedies, emergency responders and others responded quickly and worked to exhaustion. In addition to the emergency responders, bottled water companies responded immediately by donating several million bottles of water to the crews at the crash sites in New York, at the Pentagon, and in Pennsylvania. The International Bottled Water Association (IBWA 2004) reports that "within hours of the first attack, bottled water was delivered where it mattered most; to emergency personnel on the scene who required ample water to stay hydrated as they worked to rescue victims and clean up debris" (2).

Bottled water companies continued to provide bottled water to responders and rescuers at the 9/11 sites throughout the postevent(s) process(es). These patriotic actions by the bottled water companies, however, beg the question: how do we ensure the safety and security of the bottled water provided to anyone? IBWA (2004) has the answer: using a multibarrier approach, along with other principles, will enhance the safety and security of bottled water. IBWA (2004) describes its multibarrier approach as follows:

> **A multi-barrier approach**—Bottled water products are produced utilizing a multi-barrier approach, from source to finished product, that helps prevent possible harmful contaminants (physical, chemical or microbiological) from adulterating the finished product as well as storage, production, and transportation equipment. Measures in a multi-barrier approach may include source protection, source monitoring, reverse osmosis, distillation, filtration, ozonation or ultraviolet (UV) light. Many of the steps in a multi-barrier system may be effective in safeguarding bottled water from microbiological and other contamination. Piping in and out of plants, as well as storage silos and water tankers are also protected and maintained through sanitation procedures. In addition, bottled water products are bottled in a controlled, sanitary environment to prevent contamination during the filling operation. (3)

In energy sector industry infrastructure security, protection in depth is used to describe a layered security approach. A protection in depth strategy uses several forms of security techniques and/or devices against an intruder and does not rely on one single defensive mechanism to protect infrastructure. By implementing multiple layers of security, a hole or flaw in one layer is covered by the other layers. An intruder will have to intrude through each layer without being detected in the process—the layered

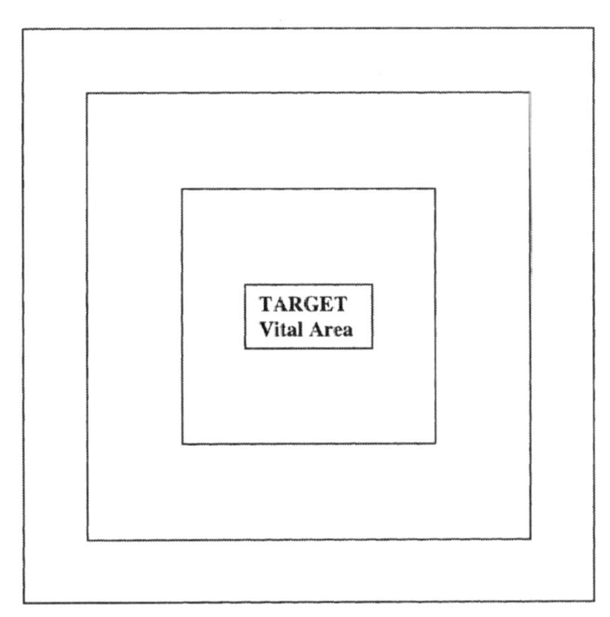

FIGURE 9.1
Layered Approach to Security

approach implies that no matter how an intruder attempts to accomplish his goal, he will encounter effective elements of the physical protection system.

For example, as depicted in figure 9.1, an effective security layering approach requires that an adversary penetrate multiple separate barriers to gain entry to a critical target at an energy facility. As shown in figure 9.1, protection in depth (multiple layers of security) helps to ensure that the security system remains effective in the event of a failure or an intruder bypassing a single layer of security.

Again, as shown in figure 9.1, layered security starts with the outer perimeter (the fence—the first line of physical security) of the facility and goes inward to the facility, the buildings, structures, other individual assets, and finally to the contents of those buildings—the targets.

The area between the outer perimeter and structures or buildings is known as the site. This open site area provides an incomparable opportunity for early identification of an unauthorized intruder and initiation of early warning/response. This open-space area is commonly used to calculate the standoff distance, that is, the distance between the outside perimeter (public areas to the fence) and the target or critical assets (buildings/structures) inside the perimeter (inside the fence line—the restricted access area).

The open area, between perimeter fence and target (e.g., operations center), if properly outfitted with various security devices, can also provide layered protection against intruders. For example, lighting is a deterrent. Based on personal experience, an open area within the plant site that is almost as well lighted at night as would be expected during daylight hours is the rule of thumb. In addition, strategically placed motion detectors along with crash barriers at perimeter gate openings and in front of vital structures are also recommended. Armed mobile guards who roam the interior of the plant site on a regular basis provide the ultimate in site area security.

The next layer of physical security is the outside wall of the target structure(s) itself. Notwithstanding door, window, and/or skylight entry, walls prevent most intruders from easy entry. If doors can be entered only by using card readers, security is shored up or enhanced to an extent. The same can be said for windows and skylights that are fashioned small enough to prohibit normal human entry. These same "weak" spots in buildings can be bastioned with break-proof or reinforced security glass.

The final layer of security is provided by properly designed interior features of buildings. Examples of these types of features include internal doors and walls, equipment cages, and backup or redundant equipment.

In the preceding discussion, the conditions described referred to perfect-world conditions; that is, to those conditions that we "would want" (i.e., the security manager's proverbial wish list) to be incorporated into the design and installation of new energy sector infrastructure. Post-9/11, in a not-so-perfect world, however, many of the peripheral (fence line) measures described above are more difficult to incorporate into energy sector infrastructure. This is not to say that energy sector sites and facilities do not have fence lines or fences; many of them do. These fences are designed to keep vandals, thieves, and trespassers out. One problem is that many of these facilities were constructed several years ago, before urban encroachment literally encircled the many sites, allowing at present little room for security stand backs or setbacks to be incorporated into electrical power stations, plants, and/or critical equipment locations. Based on personal observation, many of these fences face busy city streets or closely abut structures outside the fence line. The point is that when one sits down to plan a security upgrade, these factors must be taken into account.

The problem in protecting energy sector sites/facilities/equipment is much bigger, about one hundred times more complicated, than urban encroachment, however. Actually, the lack of urban encroachment is one of the main problems in trying to secure energy sector sites. Think about it!

The electrical sector has more than 3,300 power plants and transmission lines that extend here, there, and everywhere for more than 211,000 miles across the contiguous United States. Then there is the oil and natural gas sector. Sixty-six percent of our oil is imported at various unloading ports throughout the nation. There are 149 refineries and 116,000 miles of product pipelines. Sixteen percent of our natural gas is imported to various ports of entry. There are more than 550 gas processing plants. Gas distribution pipelines criss-cross the U.S. for more than 1,175,000 miles. The problem with protecting the energy sector is one of distance, wide dispersion, and remoteness, complicated by hackability of cyber networks and the SCADA grid control systems.

The multiple-barrier approach to security that was described earlier can be employed with relative ease to other critical infrastructures such as chemical sites, defense industrial bases, postal, public health, monuments and icons, commercial assets, and others. The same cannot be said for protecting many of the remote electrical transmission towers, many of which are 150 feet in height, with cross-arms as much as one hundred feet wide and thus the most visible component of electric transmission systems. To a limited degree, stepped-up security patrols are the partial answer to security concerns of the remote infrastructure such as electrical transmission towers and oil and natural gas pipelines.

Managers of energy sector infrastructure have four primary security areas to manage. These security areas are listed and described below.

- Physical security—In the energy sector, physical security techniques and practices have the most effective at sites such as oil refineries, natural gas processing plants, electrical substations, and other "fenced" locations. At such locations, a systems approach is best, where detection, assessment, communication, and response are planned and supported by resources, procedures, and policies.
- Cyber/information technology security—Only the use of SCADA and other key operating systems that have been properly vetted and scrubbed of alleged Chinese and/or Russian Trojan horses hacked into the North American electrical grid is important. The only positive way to ensure the security of the North American grid is to disconnect its cyber and other digital systems from the Internet. This step is impractical at the present time but points to the need to conduct frequent audits of the system and install firewall protection in SCADA and other systems to prevent hacking. Frequent third-party penetration testing is advised.
- Employment screening—Screening mitigates the threat from the enemy at the water cooler (inside the organization). We are always amazed whenever we conduct security audits for various companies. Often, a simple check such as reviewing an employee's driving record often reveals that the employee has no license, is driving on a suspended license, or has a horrific driving record. Hiring standards and pre-employment background investigations may help ensure the trustworthiness and reliability of personnel who have unescorted access to critical facilities.
- Protecting potentially sensitive information—The old saying that a secret can best be held between three people, so long as two of the three are dead, makes the point that reducing the likelihood that information could be used by those intent on disrupting operations or causing death and/or destruction in energy sector plants/sites is crucial. Information should be shared within an organization only on a need-to-know basis.

For existing facilities, security upgrades should be based on the results generated from the vulnerability assessment, which characterizes and prioritizes those assets that may be targeted. Those vulnerabilities identified must be protected.

In the following sections, various security hardware and/or devices are described. These devices serve the main purpose of providing security against physical and/or digital intrusion. That is, they are designed to delay and deny intrusion and are normally coupled with detection and assessment technology. Possible additional security measures, based on the vulnerability assessment that may be recommended (covered in this text) include the following (NAERC 2002):

- Electronic security
- Closing non-essential perimeter and internal portals
- Physical barriers such as bollards or Jersey walls
- Fencing
- Lighting
- Security surveys
- Vulnerability assessments
- Availability of security resources
- General personnel and security officer training
- Law enforcement liaison
- Ensuring availability of essential spare parts (wire, pipe, valves, transformers, etc.) for critical facilities

Keep in mind, however, and as mentioned previously, no matter the type of security device or system employed, energy sector systems cannot be made immune to all possible intrusions or attacks. Whenever a plant or facility safety/security manager tells us that he or she has secured the site 100 percent, we are reminded of Schneier's (2000) view of security: "You can't defend. You can't prevent. The only thing you can do is detect and respond." Simply, when it comes to making anything absolutely secure from intrusion or attack, there is inherently, or otherwise, no silver bullet.

SECURITY HARDWARE DEVICES*

The U.S. Environmental Protection Agency (2005) groups infrastructure security devices or products described below into four general categories:

*It is important to point out that even though the following USEPA security asset and device recommendations were first made for the water/wastewater critical infrastructure, these recommendations are applicable to all other critical infrastructure sectors, including the energy sector.

- Physical asset monitoring and control devices
- Cyber protection devices
- Communication/integration
- Environmental monitoring devices

Physical Asset Monitoring and Control Devices

Aboveground, Outdoor Equipment Enclosures

Energy sector facilities and sites usually consist of multiple components spread over a wide area (e.g., transmission towers and pipelines reaching thousands of acres and miles in some cases) and typically include a centralized distribution center, as well as oil and/or natural gas storage and distribution system components that are typically distributed at multiple locations throughout the area. Although thousands of miles of electrical cable and oil/gas pipelines are installed underground, in recent years, energy sector site/plant system designers have favored placing critical equipment—especially assets that require regular use and maintenance—aboveground.

One of the primary reasons for doing so is that locating this equipment aboveground eliminates the safety risks associated with confined-space entry, which is often required for the maintenance of equipment located belowground. In addition, space restrictions often limit the amount of equipment that can be located inside, and there are concerns that some types of equipment (such as backflow prevention devices—to prevent petrochemicals and fuel wastes from entering plant and offsite potable water systems) can, under certain circumstances, discharge fuel slurry or waste mixtures that could flood pits, vaults, or equipment rooms. In regard to electrical power, electrical substations are not usually suited for underground installation. Therefore, many pieces of critical electrical equipment are located outdoors and aboveground in configurations that are properly fenced, insulated, or isolated to prevent accidental electrical shock or short circuits/fires in equipment.

Experience demonstrates that many different system components can be and are often installed outdoors and aboveground. Examples of these types of components could include

- Backflow prevention devices
- Air release and control valves
- Pressure vacuum breakers
- Oil and gas pumps and motors
- Petrochemical storage and feed equipment
- Meters
- Sampling equipment
- Instrumentation

- Electrical substations
- Oil and natural gas pipelines

Much of this equipment is installed in remote locations and/or in areas where the public (and terrorists) can access it.

One of the most effective security measures for protecting aboveground equipment, where feasible, is to place it inside a building or exterior fenced structure. When/where this is not possible, enclosing the equipment or parts of the equipment using some sort of commercial or homemade add-on structure may help to prevent tampering with the equipment. These types of add-on structures or enclosures, which are designed to protect people and animals from electrocution and to protect equipment both from the elements and from unauthorized access or tampering, typically consist of a boxlike fenced structure that is placed over or around the entire component, or over/around critical parts of the component (i.e., valves, etc.) and is then secured to delay or prevent intruders from tampering with the equipment. The enclosures are typically locked or otherwise anchored to a solid foundation, which makes it difficult for unauthorized personnel to remove the enclosure and access the equipment.

Standardized aboveground enclosures are available in a wide variety of materials, sizes, and configurations. Many options and security features are also available for each type of enclosure, and this allows system operators the flexibility to customize an enclosure for a specific application and/or price range. In addition, most manufacturers can custom-design enclosures if standard, off-the-shelf enclosures do not meet a user's needs.

Many of these enclosures are designed to meet certain standards. For example, the American Society of Sanitary Engineers (ASSE) has developed Standard #1060, *Performance Requirements for Outdoor Enclosures for Backflow Prevention Assemblies.* If an enclosure will be used to house a backflow preventer, this standard specifies the acceptable construction materials for the enclosure as well as the performance requirements that the enclosure should meet, including specifications for freeze protection, drainage, air inlets, access for maintenance, and hinge requirements. ASSE #1060 also states that the enclosure should be lockable to enhance security.

Electrical substation and electrical equipment enclosures must meet the requirements and recommendations of various OHSA standards, the National Fire Protection Association (NFPA), the National Electrical Codes (NEC), the Institute of Electrical and Electronic Engineers (IEEE), and local code requirements.

Equipment enclosures can generally be categorized into one of four main configurations, which include

- One piece, drop-over enclosures
- Hinged or removable top enclosures

- Sectional enclosures
- Shelters with access locks

All enclosures, including those with integral floors, must be secured to a foundation to prevent them from being moved or removed. Unanchored or poorly anchored enclosures may be blown off the equipment being protected or may be defeated by intruders. In either case, this may result in the equipment beneath the enclosure becoming exposed and damaged. Therefore, ensuring that the enclosure is securely anchored will increase the security of the protected equipment.

The three basic types of foundations that can be used to anchor the aboveground equipment enclosure are concrete footers, concrete slabs on grade, or manufactured fiberglass pads. The most common types of foundations utilized for equipment enclosures are standard or slab on grade footers; however, local climate and soil conditions may dictate whether either of these types of foundations can be used. These foundations can be either precast or poured in place at the installation site. Once the foundation is installed and properly cured, the equipment enclosure is bolted or anchored to the foundation to secure it in place.

An alternative foundation, specifically for use with smaller Hot Box enclosures, is a manufactured fiberglass pad known as the Glass Pad. The Glass Pad has the center cut out so that it can be dropped directly over the piece of equipment being enclosed. Once the pad is set level on the ground, it is backfilled over a two-inch flange located around its base. The enclosure is then placed on top of the foundation and is locked in place with either a staple anchor or slotted anchor, depending on the enclosure configuration.

One of the primary attributes of a security enclosure is its strength and resistance to breaking and penetration. Accordingly, the materials from which the enclosure is constructed will be important in determining the strength of the enclosure and thus its usefulness for security applications. Enclosures are typically manufactured from either fiberglass or aluminum. With the exception of the one-piece drop-over enclosure, which is typically fabricated from fiberglass, each configuration described above can be constructed from either material. In addition, enclosures can be custom manufactured from polyurethane, galvanized steel, or stainless steel. Galvanized or stainless steel is often offered as an exterior layer, or "skin," for an aluminum enclosure. Although they are typically utilized in underground applications, precast concrete structures can also be used as aboveground equipment enclosures. However, precast structures are much heavier and more difficult to maneuver than are their fiberglass and aluminum counterparts. Concrete is also brittle, and that can be a security concern; however, products can be applied to concrete structures to add strength and minimize security risks (i.e., epoxy coating). Because precast concrete structures can be purchased from any concrete producers, this document does not identify specific vendors for these types of products.

In addition to the construction materials, enclosure walls can be configured or reinforced to give them added strength. Adding insulation is one option that can strengthen the structural characteristics of an enclosure; however, some manufacturers offer additional features to add strength to exterior walls. For example, while most enclosures are fabricated with a flat-wall construction, some vendors manufacture fiberglass shelters with ribbed exterior walls. These ribs increase the structural integrity of the wall and allow the fabrication of standard shelters up to twenty feet in length. Another vendor has developed a proprietary process that uses a series of integrated fiberglass beams that are placed throughout a foam inner core to tie together the interior and exterior walls and roof. Yet another vendor constructs aluminum enclosures with horizontal and vertical redwood beams for structural support.

Other security features that can be implemented on aboveground, outdoor equipment enclosures include locks, mounting brackets, tamper-resistant doors, and exterior lighting.

Active Security Barriers (Crash Barriers)

Active security barriers (also known as crash barriers) are large structures that are placed in roadways at entrance and exit points to protected facilities to control vehicle access to these areas. These barriers are placed perpendicular to traffic to block the roadway so that the only way that traffic can pass the barrier is for the barrier to be moved out of the roadway. These types of barriers are typically constructed from sturdy materials, such as concrete or steel, such that vehicles cannot penetrate them. They are also designed at a certain height off the roadway so that vehicles cannot go over them.

The key difference between active security barriers, which include wedges, crash beams, gates, retractable bollards, and portable barricades, and passive security barriers, which include nonmoveable bollards, Jersey barriers, and planters, is that active security barriers are designed so that they can be raised and lowered or moved out of the roadway easily to allow authorized vehicles to pass them. Many of these types of barriers are designed so that they can be opened and closed automatically (i.e., mechanized gates, hydraulic wedge barriers), while others are easy to open and close manually (swing crash beams, manual gates). In contrast to active barriers, passive barriers are permanent, nonmovable barriers, and thus they are typically used to protect the perimeter of a protected facility, such as sidewalks and other areas that do not require vehicular traffic to pass them. Several of the major types of active security barriers such as wedge barriers, crash beams, gates, bollards, and portable/removable barricades are described below.

Wedge barriers are plated, rectangular steel buttresses approximately two to three feet high that can be raised and lowered from the roadway. When they are in the open position, they are flush with the roadway, and vehicles can pass over them. However, when they are in the closed (armed) position, they project up from the road at a forty-five-degree angle, with the upper end pointing toward the oncoming vehicle and the base of the barrier away from the vehicle. Generally, wedge barriers are constructed from heavy-gauge steel or concrete that contains an impact-dampening iron rebar core that is strong and resistant to breaking or cracking, thereby allowing them to withstand the impact from a vehicle attempting to crash through them. In addition, both of these materials help to transfer the energy of the impact over the barrier's entire volume, thus helping to prevent the barrier from being sheared off its base. In addition, because the barrier is angled away from traffic, the force of any vehicle impacting the barrier is distributed over the entire surface of the barrier and is not concentrated at the base, which helps prevent the barrier from breaking off at the base. Finally, the angle of the barrier helps hang up any vehicles attempting to drive over it.

Wedge barriers can be fixed or portable. Fixed wedge barriers can be mounted on the surface of the roadway (surface-mounted wedges) or in a shallow mount in the road's surface, or they can be installed completely below the road surface. Surface-mounted wedge barricades operate by rising from a flat position on the surface of the roadway, while shallow-mount wedge barriers rise from their resting position just below the road surface. In contrast, below-surface wedge barriers operate by rising from beneath the road surface. Both the shallow-mounted and surface-mounted barriers require little or no excavation and thus do not interfere with buried utilities. All three barrier mounting types project above the road surface and block traffic when they are raised into the armed position. Once they are disarmed and lowered, they are flush with the road, thereby allowing traffic to pass. Portable wedge barriers are moved into place on wheels that are removed after the barrier has been set into place.

Installing rising wedge barriers requires preparation of the road surface. Installing surface-mounted wedges does not require that the road be excavated; however, the road surface must be intact and strong enough to allow the bolts anchoring the wedge to the road surface to attach properly. Shallow-mount and below-surface wedge barricades require excavation of a pit that is large enough to accommodate the wedge structure as well as any arming/disarming mechanisms. Generally, the bottom of the excavation pit is lined with gravel to allow for drainage. Areas not sheltered from rain or surface runoff can install a gravity drain or self-priming pump. Table 9.1 lists the pros and cons of wedge barriers.

Crash beam barriers consist of aluminum beams that can be opened or closed across the roadway. While there are several different crash beam designs, every crash beam system consists of an aluminum beam that is supported on each side by a solid footing

Table 9.1. Pros and Cons of Wedge Barriers

Pros	*Cons*
Can be surface mounted or completely installed below the roadway surface.	Installations below the surface of the roadway will require construction that may interfere with buried utilities.
Wedge barriers have a quick response time (normally 3.5–10.5 seconds, but barrier can be activated in 1–3 seconds in emergency situations. Because emergency activation of the barrier causes more wear and tear on the system than does normal activation, it is recommended for use only in true emergency situations.	Regular maintenance is needed to keep wedge fully operational.
Surface or shallow-mount wedge barricades can be utilized in locations with a high water table and/or corrosive soils.	Improper use of the system may result in authorized vehicles being hung up by the barrier and damaged. Guards must be trained to use the system properly to ensure that this does not happen. Safety technologies may also be installed to reduce the risk of the wedge activating under an authorized vehicle.
All three wedge barrier designs have a high crash rating, thereby allowing them to be employed for higher-security applications.	
These types of barriers are extremely visible, which may deter potential intruders.	

Source: U.S. Environmental Protection Agency, *Water and Wastewater Security Product Guide*, http://cfpub.epa.gov/safewater/watersecurity/guide (accessed April 4, 2009).

or buttress, which is typically constructed from concrete, steel, or some other strong material. Beams typically contain an interior steel cable (typically at least one inch in diameter) to give the beam added strength and rigidity. The beam is connected by a heavy-duty hinge or other mechanism to one of the footings so that it can swing or rotate out of the roadway when it is open and can swing back across the road when it is in the closed (armed) position, blocking the road and inhibiting access by unauthorized vehicles. The nonhinged end of the beam can be locked into its footing, thus providing anchoring for the beam on both sides of the road and increasing the beam's resistance to any vehicles attempting to penetrate it. In addition, if the crash beam is hit by a vehicle, the aluminum beam transfers the impact energy to the interior cable, which in turn transfers the impact energy through the footings and into their foundation, thereby minimizing the chance that the impact will snap the beam and allow the intruding vehicle to pass through.

Crash beam barriers can employ drop-arm, cantilever, or swing beam designs. Drop-arm crash beams operate by raising and lowering the beam vertically across the road. Cantilever crash beams are projecting structures that are opened and closed by extending the beam from the hinge buttress to the receiving buttress located on the opposite side of the road. In the swing beam design, the beam is hinged to the buttress such that it swings horizontally across the road. Generally, swing beam and cantilever designs are used at locations where a vertical lift beam is impractical. For example, the swing beam or cantilever designs are utilized at entrances and exits with overhangs, trees, or buildings that would physically block the operation of the drop-arm beam design.

Installing any of these crash beam barriers involves the excavation of a pit approximately forty-eight inches deep for both the hinge and the receiver footings. Due to the depth of excavation, the site should be inspected for underground utilities before digging begins. Table 9.2 lists the pros and cons of crash beams.

In contrast to wedge barriers and crash beams, which are typically installed separately from a fence line, *gates* are often integrated units of a perimeter fence or wall around a facility.

Gates are basically movable pieces of fencing that can be opened and closed across a road. When the gate is in the closed (armed) position, the leaves of the gate lock into steel buttresses that are embedded in concrete foundation located on both sides of the roadway, thereby blocking access to the roadway. Generally, gate barricades are constructed from a combination of heavy-gauge steel and aluminum that can absorb an impact from vehicles attempting to ram through them. Any remaining impact energy not absorbed by the gate material is transferred to the steel buttresses and their concrete foundation.

Table 9.2. Pros and Cons of Crash Beams

Pros	*Cons*
Requires little maintenance, while providing long-term durability.	Crash beams have a slower response time (normally 9.5–15.3 seconds but can be reduced to 7–10 seconds in emergency situations) than do other types of active security barriers, such as wedge barriers. Because emergency activation of the barrier causes more wear and tear on the system than does normal activation, it is recommended for use only in true emergency situations.
No excavation is required in the roadway itself to install crash beams.	All three crash beam designs possess a low crash rating relative to other types of barriers, such as wedge barriers, and thus they typically are used for lower-security applications.
	Certain crash barriers may not be visible to oncoming traffic and therefore may require additional lighting and/or other warning markings to reduce the potential for traffic to accidentally run into the beam.

Source: U.S. Environmental Protection Agency, *Water and Wastewater Security Product Guide*, http://cfpub.epa.gov.safewater/watersecurity/guide (accessed April 4, 2009).

Gates can utilize a cantilever, linear, or swing design. Cantilever gates are projecting structures that operate by extending the gate from the hinge footing across the roadway to the receiver footing. A linear gate is designed to slide across the road on tracks via a rack-and-pinion drive mechanism. Swing gates are hinged so that they can swing horizontally across the road.

Installation of the cantilever, linear, or swing gate designs described above involves the excavation of a pit approximately forty-eight inches deep for both the hinge and receiver footings to which the gates are attached. Due to the depth of excavation, the site should be inspected for underground utilities before digging begins. Table 9.3 lists the pros and cons of gates.

Bollards are vertical barriers at least three feet tall and one to two feet in diameter that are typically set four to five feet apart from each other so that they block vehicles from passing between them. Bollards can either be fixed in place, removable, or retractable. Fixed and removable bollards are passive barriers that are typically used along building perimeters or on sidewalks to prevent vehicles from them while allowing pedestrians to pass them. In contrast to passive bollards, retractable bollards are active security barriers that can easily be raised and lowered to allow vehicles to pass between them. Thus, they can be used in driveways or on roads to control vehicular access. When the bollards are raised, they protect above the road surface and block the

Table 9.3. Pros and Cons of Gates

Pros	*Cons*
All three gate designs possesses an intermediate crash rating, thereby allowing them to be utilized for medium- to higher-security applications.	Gates have a slower response time (normally 10–15 seconds, but can be reduced to 7–10 seconds in emergency situations) than do other types of active security barriers, such as wedge barriers. Because emergency activation of the barrier causes more wear and tear on the system than does normal activation, it is recommended for use only in true emergency situations.
Requires very little maintenance.	
Can be tailored to blend in with perimeter fencing.	
Gate construction requires no roadway excavation.	
Cantilever gates are useful for roads with high crowns or drainage gutters.	
These types of barriers are extremely visible, which may deter intruders.	
Gates can also be used to control pedestrian traffic.	

Source: U.S. Environmental Protection Agency, *Water and Wastewater Security Product Guide*, http://cfpub.epa.gov/safewater/watersecurity/guide (accessed April 4, 2009).

roadway; when they are lowered, they sit flush with the road surface and thus allow traffic to pass over them. Retractable bollards are typically constructed from steel or other materials that have a low weight-to-volume ratio so that they require low power to raise and lower. Steel is also more resistant to breaking than is a more brittle material, such as concrete, and is better able to withstand direct vehicular impact without breaking apart.

Retractable bollards are installed in a trench dug across a roadway—typically at an entrance or gate. Installing retractable bollards requires preparing the road surface. Depending on the vendor, bollards can be installed either in a continuous slab of concrete or in individual excavations with concrete poured in place. The required excavation for a bollard is typically slightly wider and slightly deeper than the bollard height when extended aboveground. The bottom of the excavation is typically lined with gravel to allow drainage. The bollards are then connected to a control panel, which controls the raising and lowering of the bollards. Installation typically requires mechanical, electrical, and concrete work; if utility personnel with these skills are available, then the utility can install the bollards itself. Table 9.4 lists the pros and cons of retractable bollards.

Portable/removable barriers, which can include removable crash beams and wedge barriers, are mobile obstacles that can be moved in and out of position on a roadway. For example, a crash beam may be completely removed and stored off-site when it is not needed. An additional example would be a wedge barrier that is equipped with wheels that can be removed after the barricade is towed into place.

When portable barricades are needed, they can be moved into position rapidly. To provide them with added strength and stability, they are typically anchored to buttress boxes that are located on either side of the road. These buttress boxes, which may or

Table 9.4. Pros and Cons of Retractable Bollards

Pros	*Cons*
Bollards have a quick response time (normally 3 to 10 seconds, but can be reduced to 1–3 seconds in emergency situations).	Bollard installations will require construction below the surface of the roadway, which may interfere with buried utilities.
Bollards have an intermediate crash rating, which allows them to be utilized for medium- to higher-security applications.	Some maintenance is needed to ensure barrier is free to move up and down.
	The distance between bollards must be decreased (i.e., more bollards must be installed along the same perimeter) to make these systems effective against small vehicles (i.e., motorcycles).

Source: U.S. Environmental Protection Agency, *Water and Wastewater Security Product Guide*, http://cfpub.epa.gov/safewater/watersecurity/guide (accessed April 4, 2009).

may not be permanent, are usually filled with sand, water, cement, gravel, or concrete to make them heavy and aid in stabilizing the portable barrier. In addition, these buttresses can help dissipate any impact energy from vehicles crashing into the barrier itself.

Because these barriers are not anchored into the roadway, they do not require excavation or other related construction for installation. In contrast, they can be assembled and made operational in a short period of time. The primary shortcoming of this type of design is that these barriers may move if they are hit by vehicles. Therefore, it is important to carefully assess the placement and anchoring of these types of barriers to ensure that they can withstand the types of impacts that may be anticipated at that location. Table 9.5 lists the pros and cons of portable/removable barricades.

Because the primary threat to active security barriers is that vehicles will attempt to crash through them, their most important attributes are their size, strength, and crash resistance. Other important features for an active security barrier are the mechanisms by which the barrier is raised and lowered to allow authorized vehicle entry and other factors, such as weather resistance and safety features.

Alarms

An *alarm system* is a type of electronic monitoring system that is used to detect and respond to specific types of events—such as unauthorized access to an asset or a possible fire. In chemical processing systems, alarms are also used to alert operators when process operating or monitoring conditions go out of preset parameters (i.e., process alarms). These types of alarms are primarily integrated with process monitoring and reporting systems (i.e., SCADA systems). Note that this discussion does not focus on alarm systems that are not related to a facility's processes.

Table 9.5. Pros and Cons of Portable/Removable Barricades

Pros	*Cons*
Installing portable barricades requires no foundation or roadway excavation.	Portable barriers may move slightly when hit by a vehicle, resulting in a lower crash resistance.
Can be moved in and out of position in a short period of time.	Portable barricades typically require 7.75 to 16.25 seconds to move into place, and thus they are considered to have a medium response time when compared with other active barriers.
Wedge barriers equipped with wheels can be easily towed into place.	
Minimal maintenance is needed to keep barriers fully operational.	

Source: U.S. Environmental Protection Agency, *Water and Wastewater Security Product Guide*, http://cfpub.epa.gov/safewater/watersecurity/guide (accessed April 4, 2009).

Alarm systems can be integrated with fire detection systems, intrusion detection systems (IDSs), access control systems, or closed circuit television (CCTV) systems such that these systems automatically respond when the alarm is triggered. For example, a smoke detector alarm can be set up to automatically notify the fire department when smoke is detected; or an intrusion alarm can automatically trigger cameras to turn on in a remote location so that personnel can monitor that location.

An alarm system consists of sensors that detect different types of events; an arming station that is used to turn the system on and off; a control panel that receives information, processes it, and transmits the alarm; and an annunciator that generates a visual and/or audible response to the alarm. When a sensor is tripped it sends a signal to a control panel, which triggers a visual or audible alarm and/or notifies a central monitoring station. A more complete description of each of the components of an alarm system is provided below.

Detection devices (also called *sensors*) are designed to detect a specific type of event (such as smoke, intrusion, etc.). Depending on the type of event they are designed to detect, sensors can be located inside or outside of the facility or other asset. When an event is detected, the sensors use some type of communication method (such as wireless radio transmitters, conductors, or cables) to send signals to the control panel to generate the alarm. For example, a smoke detector sends a signal to a control panel when it detects smoke.

Alarms use either normally closed (NC) or normally open (NO) electric loops, or "circuits," to generate alarm signals. These two types of circuits are discussed separately below.

In NC loops or circuits, all of the system's sensors and switches are connected in series. The contacts are "at rest" in the closed (on) position, and current continually passes through the system. However, when an event triggers the sensor, the loop is opened, breaking the flow of current through the system and triggering the alarm. NC switches are used more often than are NO switches because the alarm will be activated if the loop or circuit is broken or cut, thereby reducing the potential for circumventing the alarm. This is known as a "supervised" system.

In NO loops or circuits, all of the system's sensors and switches are connected in parallel. The contacts are "at rest" in the open (off) position, and no current passes through the system. However, when an event triggers the sensor, the loop is closed. This allows current to flow through the loop, powering the alarm. NO systems are not "supervised" because the alarm will not be activated if the loop or circuit is broken or cut. However, adding an end-of-line resistor to an NO loop will cause the system to alarm if tampering is detected.

An *arming station*, which is the main user interface with the security system, allows the user to arm (turn on), disarm (turn off), and communicate with the system. How a specific system is armed will depend on how it is used. For example, while IDSs can be armed for continuous operation (twenty-four hours/day), they are usually armed and disarmed according to the work schedule at a specific location so that personnel going about their daily activities do not set off the alarms. In contrast, fire protection systems are typically armed twenty-four hours/day.

A *control panel* receives information from the sensors and sends it to an appropriate location, such as to a central operations station or to a twenty-four-hour monitoring facility. Once the alarm signal is received at the central monitoring location, personnel monitoring for alarms can respond (such as by sending security teams to investigate or by dispatching the fire department).

An *annunciator* responds to the detection of an event by emitting a signal. This signal may be visual, audible, electronic, or a combination of these three. For example, fire alarm signals will always be connected to audible annunciators, whereas intrusion alarms may not be.

Alarms can be reported locally, remotely, or both locally and remotely. Local and remotely (centrally) reported alarms are discussed in more detail below.

A *local alarm* emits a signal at the location of the event (typically using a bell or siren). A local-only alarm emits a signal at the location of the event but does not transmit the alarm signal to any other location (i.e., it does not transmit the alarm to a central monitoring location). Typically, the purpose of a local-only alarm is to frighten away intruders and possibly to attract the attention of someone who might notify the proper authorities. Because no signal is sent to a central monitoring location, personnel can respond to a local alarm only if they are in the area and can hear and/or see the alarm signal.

Fire alarm systems must have local alarms, including both audible and visual signals. Most fire alarm signal and response requirements are codified in the National Fire Alarm Code, National Fire Protection Association (NFPA) 72. NFPA 72 discusses the application, installation, performance, and maintenance of protective signaling systems and their components. In contrast to fire alarms, which require a local signal when fire is detected, many IDSs do not have a local alert device because monitoring personnel do not wish to inform potential intruders that they have been detected. Instead, these types of systems silently alert monitoring personnel that an intrusion has been detected, thus allowing monitoring personnel to respond.

In contrast to systems that are set up to transmit local-only alarms when the sensors are triggered, systems can also be set up to transmit signals to a *central location*, such as to a control room or guard post at the utility, or to a police or fire station. Most fire/smoke alarms are set up to signal both at the location of the event and at a fire station or central monitoring station. Many insurance companies require that facilities install certified systems that include alarm communication to a central station. For example, systems certified by the Underwriters Laboratory (UL) require that the alarm be reported to a central monitoring station.

The main differences between alarm systems lie in the types of event detection devices used in different systems. *Intrusion sensors*, for example, consist of two main categories: perimeter sensors and interior (space) sensors. *Perimeter intrusion sensors* are typically applied on fences, doors, walls, windows, and so on, and are designed to detect an intruder before he/she accesses a protected asset (i.e., perimeter intrusion sensors are used to detect intruders attempting to enter through a door, window, etc.). In contrast, *interior intrusion sensors* are designed to detect an intruder who has already accessed the protected asset (i.e., interior intrusion sensors are used to detect intruders once they are already within a protected room or building). These two types of detection devices can be complementary, and they are often used together to enhance security for an asset. For example, a typical intrusion alarm system might employ a perimeter glass-break detector that protects against intruders accessing a room through a window, as well as an ultrasonic interior sensor that detects intruders that have gotten into the room without using the window. Table 9.6 lists and describes types of perimeter and interior sensors.

Fire detection/fire alarm systems consist of different types of fire detection devices and fire alarm systems available. These systems may detect fire, heat, smoke, or a combination of any of these. For example, a typical fire alarm system might consist of heat sensors, which are located throughout a facility and which detect high temperatures or a certain change in temperature over a fixed time period. A different system might be outfitted with both smoke and heat detection devices. A summary of several different types of fire/smoke/heat detection sensors is provided in table 9.7.

Once a sensor in an alarm system detects an event, it must communicate an alarm signal. The two basic types of alarm communication systems are hardwired and wireless. Hardwired systems rely on wire that is run from the control panel to each of the detection devices and annunciators. Wireless systems transmit signals from a transmitter to a receiver through the air—primarily using radio or other waves. Hardwired systems are usually lower cost, more reliable (they are not affected by terrain or environmental factors), and significantly easier to troubleshoot than are wireless systems. However, a major disadvantage of hardwired systems is that it may not be possible to hardwire all locations (for example, it may be difficult to hardwire remote locations). In addition, running wires to their required locations can be both time consuming and costly. The major advantage to using wireless systems is that they can often be installed in areas where hardwired systems are not feasible. However, wireless components can be much more expensive when compared with hardwired systems. In addition, in the

Table 9.6. Perimeter and Interior Sensors

Type of Perimeter Sensor	*Description*
Foil	Foil is a thin, fragile, lead-based metallic tape that is applied to glass windows and doors. The tape is applied to the window or door, and electric wiring connects this tape to a control panel. The tape functions as a conductor and completes the electric circuit with the control panel. When an intruder breaks the door or window, the fragile foil breaks, opening the circuit and triggering an alarm condition.
Magnetic switches (reed switches)	The most widely used perimeter sensor. They are typically used to protect doors as well as windows that can be opened (windows that cannot be opened are more typically protected by foil alarms).
Glass-break detectors	Placed on glass and sense vibrations in the glass when it is disturbed. The two most common types of glass-break detectors are shock sensors and audio discriminators.
Type of Interior Sensor	*Description*
Passive infrared (PIR)	Presently the most popular and cost-effective interior sensors. PIR detectors monitor infrared radiation (energy in the form of heat) and detect rapid changes in temperature within a protected area. Because infrared radiation is emitted by all living things, these types of sensors can be very effective.
Quad PIRs	Consist of two dual-element sensors combined in one housing. Each sensor has a separate lens and a separate processing circuitry, which allows each lens to be set up to generate a different protection pattern.
Ultrasonic detectors	Emit high-frequency sound waves and sense movement in a protected area by sensing changes in these waves. The sensor emits sound waves that stabilize and set a baseline condition in the area to be protected. Any subsequent movement within the protected area by a would-be intruder will cause a change in these waves, thus creating an alarm condition.
Microwave detectors	Emit ultrahigh-frequency radio waves, and the detector senses any changes in these waves as they are reflected throughout the protected space. Microwaves can penetrate through walls, and thus a unit placed in one location may be able to protect multiple rooms.
Dual-technology devices	Incorporate two different types of sensor technology (such as PIR and microwave technology) together in one housing. When both technologies sense an intrusion, an alarm is triggered.

Source: U.S. Environmental Protection Agency, *Water and Wastewater Security Product Guide*, http://cfpub.epa.gov/safewater/watersecurity/guide (accessed April 4, 2009).

past, it has been difficult to perform self-diagnostics on wireless systems to confirm that they are communicating properly with the controller. Presently, the majority of wireless systems incorporate supervising circuitry, which allows the subscriber to know immediately if there is a problem with the system (such as a broken detection device or a low battery) or if a protected door or window has been left open.

Table 9.7. Fire/Smoke/Heat Detection Sensors

Detector Type	*Description*
Thermal detector	Sense when temperatures exceed a set threshold (fixed temperature detectors) or when the rate of change of temperature increases over a fixed time period (rate-of-rise detectors).
Duct detector	Is located within the heating and ventilation ducts of the facility. This sensor detects the presence of smoke within the system's return or supply ducts. A sampling tube can be added to the detector to help span the width of the duct.
Smoke detectors	Sense invisible and/or visible products of combustion. The two principle types of smoke detectors are photoelectric and ionization detectors. The major differences between these devices are described below: ▪ Photoelectric smoke detectors react to visible particles of smoke. These detectors are more sensitive to the cooler smoke with large smoke particles that is typical of smoldering fires. ▪ Ionization smoke detectors are sensitive to the presence of ions produced by the chemical reactions that take place with few smoke particles, such as those typically produced by fast-burning/flaming fires.
Multisensor detectors	Are a combination of photoelectric and thermal detectors. The photoelectric sensor serves to detect smoldering fires, while the thermal detector senses the heat given off from fast-burning/flaming fires.
Carbon monoxide (CO) detectors	Are used to indicate the outbreak of fire by sensing the level of carbon monoxide in the air. The detector has an electrochemical cell that senses carbon monoxide but not some other products of combustion.
Beam detectors	Are designed to protect large, open spaces such as industrial warehouses. These detectors consist of three parts: the transmitter, which projects a beam of infrared light; the receiver, which registers the light and produces an electrical signal; and the interface, which processes the signal and generates an alarm of fault signals. In the event of a fire, smoke particles obstruct the beam of light. Once a preset threshold is exceeded, the detector will go into alarm.
Flame detectors	Sense either ultraviolet (UV) or infrared (IR) radiation emitted by a fire.
Air-sampling detectors	Actively and continuously sample the air from a protected space and are able to sense the precombustion stages of incipient fire.

Source: U.S. Environmental Protection Agency, *Water and Wastewater Security Product Guide*, http://cfpub.epa.gov/safewater/watersecurity/guide (accessed April 4, 2009).

Backflow Prevention Devices

Backflow prevention devices are designed to prevent backflow, which is the reversal of the normal and intended direction of water flow in a water system. Backflow is a potential problem in a petrochemical processing system because if incorrectly cross-connected to potable water, it can spread contaminated water back through a distribution system. For example, backflow at uncontrolled cross-connections (cross-connections are any actual or potential connection between the public water supply and a source of chemical contamination) or pollution can allow pollutants or contaminants to enter the potable water system. More specifically, backflow from private plumbing systems, industrial areas, hospitals, and other hazardous contaminant-containing systems into public water mains and wells poses serious public health risks and security problems. Cross-contamination from private plumbing systems can contain biological hazards (such as bacteria or viruses) or toxic substances that can contaminate and sicken an entire population in the event of backflow. The majority of historical incidences of backflow have been accidental, but growing concern that contaminants could be intentionally be backfed into a system is prompting increased awareness for private homes, businesses, industries, and areas most vulnerable to intentional strikes. Therefore, backflow prevention is a major tool for the protection of water systems.

Backflow may occur under two types of conditions: backpressure and backsiphonage. *Backpressure* is the reverse from normal flow direction within a piping system that is the result of the downstream pressure being higher than the supply pressure. These reductions in the supply pressure occur whenever the amount of water being used exceeds the amount of water supplied, such as during water line flushing, firefighting, or breaks in water mains. *Backsiphonage* is the reverse from normal flow direction within a piping system that is caused by negative pressure in the supply piping (i.e., the reversal of normal flow in a system caused by a vacuum or partial vacuum within the water supply piping). Backsiphonage can occur where there is a high velocity in a pipe line; when there is a line repair or break that is lower than a service point; or when there is lowered main pressure due to high water withdrawal rate, such as during firefighting or water main flushing.

To prevent backflow, various types of backflow preventers are appropriate for use. The primary types of backflow preventers are

- Air gap drains
- Double-check valves
- Reduced pressure principle assemblies
- Pressure vacuum breakers

Biometric Security Systems

Biometrics involves measuring the unique physical characteristics or traits of the human body. Any aspect of the body that is measurably different from person to person—for example, fingerprints or eye characteristics—can serve as a unique biometric identifier for that individual. Biometric systems recognizing fingerprints, palm shape, eyes, face, voice, and signature comprise the bulk of the current biometric systems. However, some systems that recognize other biological features do exist.

Biometric security systems use biometric technology combined with some type of locking mechanisms to control access to specific assets. In order to access an asset controlled by a biometric security system, an individual's biometric trait must be matched with an existing profile stored in a database. If there is a match between the two, the locking mechanism (which could be a physical lock, such as at a doorway, an electronic lock, such as at a computer terminal, or some other type of lock) is disengaged, and the individual is given access to the asset.

A biometric security system typically comprises the following components:

- A sensor, which measures/records a biometric characteristic or trait
- A control panel, which serves as the connection point between various system components. The control panel communicates information back and forth between the sensor and the host computer and controls access to the asset by engaging or disengaging the system lock based on internal logic and information from the host computer
- A host computer, which processes and stores the biometric trait in a database
- Specialized software, which compares an individual image taken by the sensor with a stored profile or profiles
- A locking mechanism, which is controlled by the biometric system
- A power source to power the system

Biometric Hand and Finger Geometry Recognition

Hand and finger geometry recognition is the process of identifying an individual through the unique geometry (shape, thickness, length, width, etc.) of that individual's hand or fingers. Hand geometry recognition has been employed since the early 1980s and is among the most widely used biometric technologies for controlling access to important assets. It is easy to install and use and is appropriate for use in any location requiring use of two-finger, highly accurate, nonintrusion biometric security. For example, it is currently used in numerous workplaces, day care facilities, hospitals, universities, airports, refineries, and power plants.

A newer option within hand geometry recognition technology is finger geometry recognition (not to be confused with fingerprint recognition). Finger geometry recognition relies on the same scanning methods and technologies as does hand geometry recognition, but the scanner scans only two of the user's fingers, as opposed to the entire hand. Finger geometry recognition has been in commercial use since the mid 1990s and is mainly used in time and attendance applications (i.e., to track when individuals have entered and exited a location). To date the only large-scale commercial use of two-finger geometry for controlling access is at Disney World, where season pass holders use the geometry of their index and middle finger to gain access to the facilities.

To use a hand or finger geometry unit, an individual presents his or her hand or fingers to the biometric unit for scanning. The scanner consists of a charged coupled device (CCD), which is essentially a high-resolution digital camera; a reflective platen on which the hand is placed; and a mirror or mirrors that help capture different angles of the hand or fingers. The camera scans individual geometric characteristics of the hand or fingers by taking multiple images while the user's hand rests on the reflective platen. The camera also captures depth, or three-dimensional information, through light reflected from the mirrors and the reflective platen. This live image is then compared with a template that was previously established for that individual when he or she was enrolled in the system. If the live scan of the individual matches the stored template, the individual is verified and is given access to that asset. Typically, verification takes about two seconds. In access control applications, the scanner is usually connected to some sort of electronic lock, which unlocks the door, turnstile, or other entry barrier when the user is verified. The user can then proceed through the entrance. In time and attendance applications, the time that an individual checks in and out of a location is stored for later use.

As discussed above, hand and finger geometry recognition systems can be used in several different types of applications, including access control and time and attendance tracking. While time and attendance tracking can be used for security, it is primarily used for operations and payroll purposes (i.e., clocking in and clocking out). In contrast, access control applications are more likely to be security related. Biometric systems are widely used for access control and can be used on various types of assets, including entryways, computers, vehicles, and so on. However, because of their size, hand/finger recognition systems are primarily used in entryway access control applications.

Biometric Iris Recognition

The iris, which is the colored or pigmented area of the eye surrounded by the sclera (the white portion of the eye), is a muscular membrane that controls the amount of light entering the eye by contracting or expanding the pupil (the dark center of the eye). The dense, unique patterns of connective tissue in the human iris were first noted in 1936, but it was not unitl1994, when algorithms for iris recognition were created and patented, that commercial applications using biometric iris recognition began to be used extensively. There are now two vendors producing iris recognition technology: both the original developer of these algorithms as well as a second company, which has developed and patented a different set of algorithms for iris recognition.

The iris is an ideal characteristic for identifying individuals because it is formed *in utero*, and its unique patterns stabilize around eight months after birth. No two irises are alike, neither an individual's right or left irises nor the irises of identical twins. The iris is protected by the cornea (the clear covering over the eye), and therefore it is not subject to the aging or physical changes (and potential variation) that are common to some other biometric measures, such as the hand, fingerprints, and the face. Although some limited changes can occur naturally over time, these changes generally occur in the iris's melanin and therefore affect only the eye's color, not its unique patterns (in addition, because iris scanning uses only black-and-white images, color changes would not affect the scan anyway). Thus, barring specific injuries or certain surgeries directly affecting the iris, the iris's unique patterns remain relatively unchanged over an individual's lifetime.

Iris recognition systems employ a monochromatic or black-and-white video camera that uses both visible and near-infrared light to take video of an individual's iris. Video is used rather than still photography as an extra security procedure. The video is used to confirm the normal continuous fluctuations of the pupil as the eye focuses, which ensures that the scan is of a living human being and not a photograph or some other attempted hoax. A high-resolution image of the iris is then captured or extracted from the video, using a device often referred to as a frame grabber. The unique characteristics identified in this image are then converted into a numeric code, which is stored as a template for that user.

Card Identification/Access/Tracking Systems

A card reader system is a type of electronic identification system that is used to identify a card and then perform an action associated with that card. Depending on the system, the card may identify where a person is or was at a certain time; or it may authorize another action, such as disengaging a lock. For example, a security guard may use his card at card readers located throughout a facility to indicate that he has checked a certain location at a certain time. The reader will store the information and/or send it to a central location, where it can be checked later to ensure that the guard has patrolled the area. Other card reader systems can be associated with a lock so that the cardholder must have his or her card read and accepted by the reader before the lock disengages.

A complete card reader system typically consists of the following components:

- Access cards that are carried by the user
- Card readers, which read the card signals and send the information to control units
- Control units, which control the response of the card reader to the card
- A power source

A card may be a typical card or another type of device, such as a key fob or wand. These cards store electronic information, which can range from a simple code (i.e., the alphanumeric code on a proximity card) to individualized personal data (i.e., biometric data on a smart card). The card reader reads the information stored on the card and sends it to the control unit, which determines the appropriate action to take when a card is presented. For example, in a card access system, the control unit compares the information on the card versus stored access authorization information to determine whether the card holder is authorized to proceed through the door. If the information stored in the card reader system indicates that the key is authorized to allow entrance through the doorway, the system disengages the lock, and the key holder can proceed through the door.

There are many different types of card reader systems on the market. The primary differences between card reader systems are in the way that data is encoded on the cards, the way these data are transferred between the card and the card reader, and the types of applications for which they are best suited. However, all card systems are similar in the way that the card reader and control unit interact to respond to the card.

While card readers are similar in the way that the card reader and control unit interact to control access, they are different in the way data is encoded on the cards and the way these data are transferred between the card and the card reader. There are several types of technologies available for card reader systems. These include

- Proximity
- Wiegand
- Smart card
- Magnetic stripe
- Bar code
- Infrared
- Barium ferrite
- Hollerith
- Mixed technologies

Table 9.8 summarizes various aspects of card reader technologies. The determination for the level of security rate (low, moderate, or high) is based on the level of technology a given card reader system has and how simple it is to duplicate that technology and thus bypass the security. Vulnerability ratings were based on whether the card reader can be damaged easily due to frequent use or difficult working conditions (i.e., weather conditions if the reader is located outside). Often this is influenced by the number of moving parts in the system—the more moving parts, the greater the system's potential susceptibility to damage. The life cycle rating is based on the

Table 9.8. Card Reader Technology

Types of Card Readers	*Technology*	*Life Cycle*	*Vulnerability*	*Level of Security*
Proximity	Embedded radio frequency circuits encoded with unique information	Long	Virtually none	Moderate–high
Wiegand	Short lengths of small-diameter, special alloy wire with unique magnetic properties	Long	Low susceptibility to damage; high durability due to embedded wires	Moderate-expensive
Magnetic Stripe	Electromagnetic charges to encode information on a piece of tape attached to back of card	Moderate	Moderately susceptible to damage due to frequency of use	Low–moderate
Bar Code	Series of narrow and wide bars and spaces	Short	High; easily damaged	Low
Hollerith	Holes punched in a plastic or paper card and read optically	Short	High; easily damaged from frequent use	Low
Infrared	An encoded shadow pattern within the card, read using an infrared scanner	Moderate	IR scanners are optical and thus vulnerable to contamination	High
Barium Ferrite	Uses small bits of magnetized barium ferrite, placed inside a plastic card; the and location of polarity the "spots" determine the coding	Moderate	Low susceptibility o damage; durable since spots are embedded in the material	Moderate–high
Smartcards	Patterns or series of narrow and wide bars and spaces	Short	High susceptibility to damage, low durability	Highest

Source: U.S. Environmental Protection Agency, *Water and Wastewater Security Product Guide*, http://cfpub.epa.gov.safewater/watersecurity/guide (accessed April 4, 2009).

durability of a given card reader system over its entire operational period. Systems requiring frequent physical contact between the reader and the card often have a shorter life cycle due to the wear and tear to which the equipment is exposed. For many card reader systems, the vulnerability rating and life cycle ratings have a reciprocal relationship. For instance, if a given system has a high vulnerability rating, it will almost always have a shorter life cycle.

Card reader technology can be implemented for facilities of any size and with any number of users. However, because individual systems vary in the complexity of their technology and in the level of security they can provide to a facility, individual users must determine the appropriate system for their needs. Some important features to consider when selecting a card reader system include

- The technological sophistication and security level of the card system.
- The size and security needs of the facility.
- The frequency with which the card system will be used. For systems that will experience a high frequency of use, it is important to consider a system that has a longer life cycle and lower vulnerability rating, thus making it more cost effective to implement.
- The conditions in which the system will be used (i.e., will it be used on the interior or exterior of buildings, does it require light or humidity controls, etc.). Most card reader systems can operate under normal environmental conditions; therefore, this would be a mitigating factor only in extreme conditions.
- System costs.

Exterior Intrusion Sensors

An exterior intrusion sensor is a detection device that is used in an outdoor environment to detect intrusions into a protected area. These devices are designed to detect an intruder and then communicate an alarm signal to an alarm system. The alarm system can respond to the intrusion in many different ways, such as by triggering an audible or visual alarm signal or by sending an electronic signal to a central monitoring location that notifies security personnel of the intrusion.

Intrusion sensors can be used to protect many kinds of assets. Intrusion sensors that protect physical space are classified according to whether they protect indoor, or interior space (i.e., an entire building or room within a building), or outdoor, or exterior space (i.e., a fence line or perimeter). Interior intrusion sensors are designed to protect the interior space of a facility by detecting an intruder who is attempting to enter or who has already entered a room or building. In contrast, exterior intrusion sensors are designed to detect an intrusion into a protected outdoor/exterior area. Exterior protected areas are typically arranged as zones or exclusion areas placed so that the intruder is detected early in the intrusion attempt before the intruder can gain access to more valuable assets (e.g., a building located within the protected area). Early detection creates additional time for security forces to respond to the alarm.

Exterior intrusion sensors are classified according to how the sensor detects the intrusion within the protected area. The three classes of exterior sensor technology include

- Buried-line sensors
- Fence-associated sensors
- Freestanding sensors

1. Buried-line sensors—As the name suggests, buried-line sensors are sensors that are buried underground and are designed to detect disturbances within the ground—such as disturbances caused by an intruder digging, crawling, walking, or running on the monitored ground. Because they sense ground disturbances, these types of sensors are able to detect intruder activity both on the surface and belowground. Individual types of exterior buried-line sensors function in different ways, including, by detecting motion, pressure, or vibrations within the protected ground, or by detecting changes in some type of field (e.g., magnetic field) that the sensors generate within the protected ground. Specific types of buried line sensors include pressure or seismic sensors, magnetic field sensors, ported coaxial cables, and fiber-optic cables. Details on each of these sensor types are provided below. Table 9.9 presents the distinctions among the four types of buried sensors.

   - *Buried-line pressure* or *seismic sensors* detect physical disturbances to the ground—such as vibrations or soil compression—caused by intruders walking, driving, digging, or otherwise physically contacting the protected ground. These sensors detect disturbances from all directions and, therefore, can protect an area radially outward from their location; however, because detection may weaken as

Table 9.9. Types of Buried Sensors

Type	*Description*
Pressure or seismic	Responds to disturbances in the soil.
Magnetic field	Responds to a change in the local magnetic field caused by the movement of nearby metallic material.
Ported coaxial cables	Respond to motion of a material with a high dielectric constant or high conductivity near the cables.
Fiber-Optic Cables	Respond to a change in the shape of the fiber that can be sensed using sophisticated sensors and computer signal processing.

Source: Adapted from M. L. Garcia, *The Design and Evaluation of Physical Protection Systems* (Burlington, MA: Butterworth-Heinemann, 2001).

a function of distance from the disturbance, choosing the correct burial depth in the area to be protected will be crucial. In general, sensors buried at a shallow depth protect a relatively small area but have a high probability of detecting intrusion within that area, while sensors buried at a deeper depth protect a wider area but have a lower probability of detecting intrusion into that area.

- *Buried-line magnetic field sensors* detect changes in a local magnetic field that are caused by the movement of metallic objects within that field. This type of sensor can detect ferric metal objects worn or carried by an intruder entering a protected area on foot as well as vehicles being driven into the protected area.
- *Buried-line ported coaxial cable sensors* detect the motion of any object (i.e., human body, metal, etc.) possessing high conductivity and located within close proximity to the cables. An intruder entering into the protected space creates an active disturbance in the electric field, thereby triggering an alarm condition.
- *Buried-line fiber-optic cable sensors* detect changes in the attenuation of light signals transmitted within the cable. When the soil around the cable is compressed, the cable is distorted, and the light signal transmitted through the cable changes, initiating an alarm. This type of sensor is easy to install because it can be buried at a shallow burial depth (only a few centimeters) and still be effective.

2. Fence-associated sensors—Fence-associated sensors are either attached to an existing fence or are installed in such a way as to create a fence. These sensors detect disturbances to the fence—such as those caused by an intruder attempting to climb the fence, or by an intruder attempting to cut or lift the fence fabric. Exterior fence-associated sensors include fence-disturbance sensors, taut-wire sensor fences, and electric field or capacitance sensors. Details on each of these sensor types are provided below.
   - *Fence-disturbance sensors* detect the motion or vibration of a fence, such as that caused by an intruder attempting to climb or cut through the fence. In general, fence disturbance sensors are used on chain link fences or on other fence types where a movable fence fabric is hung between fence posts.
   - *Taut-wire sensor fences* are similar to fence-disturbance sensors except that instead of attaching the sensors to a loose fence fabric, the sensors are attached to a wire that is stretched tightly across the fence. These types of systems are designed to detect changes in the tension of the wire rather than vibrations in the fence fabric. Taut-wire sensor fences can be installed over existing fences or as standalone fence systems.
   - *Electric field or capacitance sensors* detect changes in capacitive coupling between wires that are attached to, but electrically isolated from, the fence. As opposed to other fence-associated intrusion sensors, both electric field and

capacitance sensors generate an electric field that radiates out from the fence line, resulting in an expanded zone of protection relative to other fence-associated sensors and allowing the sensor to detect intruders' presence before they arrive at the fence line. Note: Proper spacing is necessary during installation of the electric field sensor to detect a would-be intruder from slipping between largely spaced wires.

3. *Free-standing sensors*—These sensors, which include active infrared, passive infrared, bistatic microwave, monostatic microwave, dual-technology, and video motion detection (VMD) sensors, consist of individual sensor units or components that can be set up in a variety of configurations to meet a user's needs. They are installed aboveground, and depending on how they are oriented relative to each other, they can be used to establish a protected perimeter or a protected space. More details on each of these sensor types are provided below.
   - *Active infrared sensors* transmit infrared energy into the protected space and monitor for changes in this energy caused by intruders entering that space. In a typical application, an infrared light beam is transmitted from a transmitter unit to a receiver unit. If an intruder crosses the beam, the beam is blocked, and the receiver unit detects a change in the amount of light received, triggering an alarm. Different sensors can see single- and multiple-beam arrays. Single-beam infrared sensors transmit a single infrared beam. In contrast, multiple-beam infrared sensors transmit two or more beams parallel to each other. This multiple-beam sensor arrangement creates an infrared "fence."
   - *Passive infrared (PIR) sensors* monitor the ambient infrared energy in a protected area and evaluate changes in that ambient energy that may be caused by intruders moving through the protected area. Detection ranges can exceed one hundred yards on cold days, with size and distance limitations dependent upon the background temperature. PIR sensors generate a nonuniform detection pattern (or "curtain") that has areas (or "zones") of more sensitivity and areas of less sensitivity. The specific shape of the protected area is determined by the detector's lenses. The general shape common to many detection patterns is a series of long "fingers" emanating from the PIR and spreading in various directions. When intruders enter the detection area, the PIR sensor detects differences in temperature due to the intruder's body heat and triggers an alarm. While the PIR leaves unprotected areas between its fingers, an intruder would be detected if he passed from a nonprotected area to a protected area.
   - *Microwave sensors* detect changes in received energy generated by the motion of an intruder entering into a protected area. Monostatic microwave sensors incorporate transmitter and receiver in one unit, while bistatic sensors separate the

transmitter and the receiver into different units. Monostatic sensors are limited to a coverage area of four hundred feet, while bistatic sensors can cover an area up to 1,500 feet. For bistatic sensors, a zone of no detection exists in the first few feet in front of the antennas. This distance from the antennas to the point at which the intruder is first detected is known as the offset distance. Due to this offset distance, antennas must be configured so that they overlap each other (as opposed to being adjacent to each other), thereby creating long perimeters with a continuous line of detection.

- *Dual-technology sensors* consist of two different sensor technologies incorporated together into one sensor unit. For example, a dual-technology sensor could consist of a passive infrared detector and a monostatic microwave sensor integrated into the same sensor unit.
- *Video motion detection* (VMD) *sensors* monitor video images from a protected area for changes in the images. Video cameras are used to detect unauthorized intrusion into the protected area by comparing the most recent image against a previously established one. Cameras can be installed on towers or other tall structures so that they can monitor a large area.

Fences

A fence is a physical barrier that can be set up around the perimeter of an asset. Fences often consist of individual pieces (such as individual pickets in a wooden fence or individual sections of a wrought iron fence) that are fastened together. Individual sections of the fence are fastened together using posts, which are sunk into the ground to provide stability and strength for the sections of the fence hung between them. Gates are installed between individual sections of the fence to allow access inside the fenced area.

Many fences are used as decorative architectural features to separate physical spaces from each other. They may also be used to physically mark the location of a boundary (such as a fence installed along a properly line). However, a fence can also serve as an effective means for physically delaying intruders from gaining access to an energy sector asset. For example, many utilities install fences around their primary facilities, around remote pump stations, or around hazardous petrochemical materials storage areas or sensitive areas within a facility. Access to the area can be controlled through security at gates or doors through the fence (for example, by posting a guard at the gate or by locking it). In order to gain access to the asset, unauthorized persons would either have to go around or through the fence.

Fences are often compared with walls when determining the appropriate system for perimeter security. While both fences and walls can provide adequate perimeter security, fences are often easier and less expensive to install than walls. However, they do not usually provide the same physical strength that walls do. In addition, many types of fences have gaps between the individual pieces that make up the fence (i.e., the spaces between chain links in a chain link fence or the spaces between pickets in a picket fence). Thus, many types of fences allow the interior of the fenced area to be seen. This may allow intruders to gather important information about the locations or defenses of vulnerable areas within the facility.

There are numerous types of materials used to construct fences, including chain link iron, aluminum, wood, or wire. Some types of fences, such as split rails or pickets, may not be appropriate for security purposes because they are traditionally low fences, and they are not physically strong. Potential intruders may be able to easily defeat these fences either by jumping or climbing over them or by breaking through them. For example, the rails in a split fence may be able to be broken easily.

Important security attributes of a fence include the height to which it can be constructed, the strength of the material composing the fence, the method and strength of attaching the individual sections of the fence together at the posts, and the fence's ability to restrict the view of the assets inside the fence. Additional considerations should include the ease of installing the fence and the ease of removing and reusing sections of the fence. Table 9.10 provides a comparison of the important security and usability features of various fence types.

Some fences can include additional measures to delay, or even detect, potential intruders. Such measures may include the addition of barbed wire, razor wire, or other deterrents at the top of the fence. Barbed wire is sometimes employed at the base of fences as well. This can impede a would-be intruder's progress in even reaching the fence. Fences may also be fitted with security cameras to provide visual surveillance of the perimeter. Finally, some facilities have installed motion sensors along their fences to detect movement on the fence. Several manufacturers have combined these multiple perimeter security features into one product and offer alarms and other security features.

Table 9.10. Comparison of Different Fence Types

Specifications	*Chain Link*	*Iron*	*Wire (Wirewall)*	*Wood*
Height limitations	12'	12'	12'	8'
Strength	Medium	High	High	Low
Installation requirements	Low	High	High	Low
Ability to remove/reuse	Low	High	Low	High
Ability to replace/repair	Medium	High	Low	High

Source: U.S. Environmental Protection Agency, *Water and Wastewater Security Product Guide*, http://cfpub.epa.gov/safewater/watersecurity/guide (accessed April 4, 2009).

The correct implementation of a fence can make it a much more effective security measure. Security experts recommend the following when a facility constructs a fence:

- The fence should be at least seven to nine feet high.
- Any outriggers, such as barbed wire, that are affixed on top of the fence should be angled out and away from the facility, and not in toward the facility. This will make climbing the fence more difficult and will prevent ladders from being placed against the fence.
- Other types of hardware can increase the security of the fence. This can include installing concertina wire along the fence (this can be done in front of the fence or at the top of the fence) or adding intrusion sensors, camera, or other hardware to the fence.
- All undergrowth should be cleared for several feet (typically six feet) on both sides of the fence. This will allow for a clearer view of the fence by any patrols in the area.
- Any trees with limbs or branches hanging over the fence should be trimmed so that intruders cannot use them to go over the fence. Also, it should be noted that fallen trees can damage fences, and so management of trees around the fence can be important. This can be especially important in areas where fencing goes through a remote area.
- Fences that do not block the view from outside the fence to inside the fence allow patrols to see inside the fence without having to enter the facility.
- "No Trespassing" signs posted along a fence can be a valuable tool in prosecuting any intruders who claim that the fence was broken and that they did not enter through the fence illegally. Adding signs that highlight the local ordinances against trespassing can further deter simple troublemakers from illegally jumping/climbing the fence. Electrical substation and other electrical component installations should have clearly visible signage warning of high voltage and the dangers of electrical shock.

Films for Glass Shatter Protection

Most energy sector utilities have numerous windows on the outside of buildings, in doors, and in interior offices. In addition, many facilities have glass doors or other glass structures, such as glass walls or display cases. These glass objects are potentially vulnerable to shattering when heavy objects are thrown or launched at them, when explosions occur near them, or when there are high winds (for exterior glass). If the glass is shattered, intruders may potentially enter an area. In addition, shattered glass projected into a room from an explosion or from an object being thrown through a door or window can injure and potentially incapacitate personnel in the room. Materials that prevent glass from shattering can help to maintain the integrity of the door, window, or other glass object, and can delay an intruder from gaining access. These materials can also prevent flying glass and thus reduce potential injuries.

Materials designed to prevent glass from shattering include specialized films and coatings. These materials can be applied to existing glass objects to improve their strength and their ability to resist shattering. The films have been tested against many scenarios that could result in glass breakage, including penetration by blunt objects, bullets, high winds, and simulated explosions. Thus, the films are tested against both simulated weather scenarios (which could include both the high winds themselves and the force of objects blown into the glass) as well as more criminal/terrorist scenarios where the glass is subject to explosives or bullets. Many vendors provide information on the results of these types of tests, and thus potential users can compare different product lines to determine which products best suit their needs.

The primary attributes of films for shatter protection are

- The materials from which the film is made
- The adhesive that bonds the film to the glass surface
- The thickness of the film

Standard glass safety films are designed from high-strength polyester. Polyester provides both strength and elasticity, which is important in absorbing the impact of an object, spreading the force of the impact over the entire film, and resisting tearing. The polyester is also designed to be resistant to scratching, which can result when films are cleaned with abrasives or other industrial cleaners.

The bonding adhesive is important in ensuring that the film does not tear away from the glass surface. This can be especially important when the glass is broken so that the film does not peel off the glass and allow it to shatter. In addition, films applied to exterior windows can be subject to high concentrations of UV light, which can break down bonding materials.

Film thickness is measured in gauge or mils. According to test results reported by several manufacturers, film thickness appears to affect resistance to penetration/tearing, with thicker films being more resistant to penetration and tearing. However, the application of a thicker film did not decrease glass fragmentation.

Many manufacturers offer films in different thicknesses. The "standard" film is usually one four-mil layer; thicker films are typically composed of several layers of the standard four-mil sheet. However, newer technologies have allowed the polyester to be "microlayered" to produce a stronger film without significantly increasing its thickness. In this microlayering process, each laminate film is composed of multiple microthin layers of polyester woven together at alternating angles. This provides increased strength for the film while maintaining the flexibility and thin profile of one film layer.

As described above, many vendors test their products in various scenarios that would lead to glass shattering, including simulated bomb blasts and simulation of the glass being struck by windblown debris. Some manufacturers refer to the Government Services Administration standard for bomb blasts, which require resistance to tearing for a four-PSI blast. Other manufacturers use other measures and test for resistance to tearing. Many of these tests are not standard, in that no standard testing or reporting methods have been adopted by any of the accepted standards-setting institutions. However, many of the vendors publish the procedure and the results of these tests on their websites, and this may allow users to evaluate the protectiveness of these films. For example, several vendors evaluate the "protectiveness" of their films and the "hazard" resulting from blasts near windows with and without protective films. Protectiveness is usually evaluated based on the percentage of glass ejected from the window and the height at which that ejected glass travels during the blast (for example, if the blasted glass tends to project upward into a room—potentially toward people's faces—it is a higher hazard than if it is blown downward into the room toward people's feet). There are some standard measures of glass breakage. For example, several vendors indicated that their products exceed the American Society for Testing and Materials (ASTM) Standard 64Z-95 "Standard Test Method for Glazing and Glazing Systems Subject to Air Blast Loadings." Vendors often compare the results of some sort of penetration or force test, ballistic tests, or simulated explosions with unprotected glass versus glass onto which their films have been applied. Results generally show that applying films to the glass surfaces reduces breakage/penetration of the glass and can reduce the amount and direction of glass ejected from the frame. This in turn reduces the hazard from flying glass.

In addition to these types of tests, many vendors conduct standard physical tests on their products, such as tests for tensile strength and peel strength. Tensile strength indicates the strength per area of material, while the peel strength indicates the force it would take to peel the product from the glass surface. Several vendors indicate that their products exceed American National Standards Institute (ANSI) Standard Z97.1 for tensile strength and adhesion.

Vendors typically have a warranty against peeling or other forms of deterioration of their products. However, the warranty requires that the films be installed by manufacturer-certified technicians to ensure that they are applied correctly and therefore that the warranty is in effect. Warranties from different manufacturers may vary. Some may cover the cost of replacing the material only, while others include material plus installation. Because installation costs are significantly greater than material costs, different warranties may represent large differences in potential costs.

Fire Hydrant Locks

Fire hydrants are installed at strategic locations throughout a community's water distribution system to supply water for firefighting. However, because there are many hydrants in a system and they are often located in residential neighborhoods, industrial districts, and other areas where they cannot be easily observed and/or guarded, they are potentially vulnerable to unauthorized access. Many municipalities, states, and EPA regions have recognized this potential vulnerability and have instituted programs to lock hydrants. For example, EPA Region 1 has included locking hydrants as number 7 on its "Drinking Water Security and Emergency Preparedness" top-ten list for small groundwater suppliers.

A hydrant lock is a physical security device designed to prevent unauthorized access to the water supply through a hydrant. It can also ensure water and water pressure availability to firefighters and prevent water theft and associated lost water revenue. These locks have been successfully used in numerous municipalities and in various climates and weather conditions.

Fire hydrant locks are basically steel covers or caps that are locked in place over the operating nut of a fire hydrant. The lock prevents unauthorized persons from accessing the operating nut and opening the fire hydrant valve. The lock also makes it more difficult to remove the bolts from the hydrant and access the system that way. Finally, hydrant locks shield the valve from being broken off. Should a vandal attempt to breach the hydrant lock by force and succeed in breaking the hydrant lock, the vandal will only succeed in bending the operating valve. If the hydrant's operating valve is bent, the hydrant will not be operational, but the water asset remains protected and inaccessible to vandals. However, the entire hydrant will need to be replaced.

Hydrant locks are designed so that the hydrants can be operated by special "key wrenches" without removing the lock. These specialized wrenches are generally distributed to the fire department, public works department, and other authorized persons so that they can access the hydrants as needed. An inventory of wrenches and their serial numbers is generally kept by a municipality so that the location of all wrenches is known. These operating key wrenches may only be purchased by registered lock owners.

The most important features of hydrant locks are their strength and the security of their locking systems. The locks must be strong so that they cannot be broken off. Hydrant locks are constructed from stainless or alloyed steel. Stainless steel locks are stronger and are ideal for all climates; however, they are more expensive than alloy locks. The locking mechanisms for each fire hydrant locking system ensure that the hydrant can only be operated by authorized personnel who have the specialized key to work the hydrant.

Hatch Security

A hatch is basically a door that is installed on a horizontal plane (such as in a floor, a paved lot, or a ceiling) instead of on a vertical plane (such as in a building wall). Hatches are usually used to provide access to assets that are either located underground (such as hatches to basements or underground vaults and storage areas) or to assets located above ceilings (such as emergency roof exits). At chemical industrial facilities, hatches are typically used to provide access to underground vaults containing pumps, meter chambers, valves, or piping, or to the interior of chemical tanks or covered water reservoirs. Securing a hatch by locking it or upgrading materials to give the hatch added strength can help to delay unauthorized access to any asset behind the hatch.

Like all doors, a hatch consists of a frame anchored to the horizontal structure, a door or doors, hinges connecting the door/doors to the frame, and a latching or locking mechanism that keeps the hatch door/doors closed.

It should be noted that improving hatch security is straightforward and that hatches with upgraded security features can be installed new or they can be retrofit for existing applications.

Depending on the application, the primary security-related attributes of a hatch are the strength of the door and frame, its resistance to the elements and corrosion, it ability to be sealed against water or gas, and its locking features.

Hatches must be both strong and lightweight so that they can withstand typical static loads (such as people or vehicles walking or driving over them) while still being easy to open.

In addition, because hatches are typically installed at outdoor locations, they are usually designed from corrosion-resistant metal that can withstand the elements. Therefore, hatches are typically constructed from high-gauge steel or lightweight aluminum.

Aluminum is typically the material of choice for hatches because it is lightweight and more corrosion resistant relative to steel. Aluminum is not as rigid as steel, so aluminum hatch doors may be reinforced with aluminum stiffeners to provide extra strength and rigidity. The doors are usually constructed from single or double layers (or "leaves") of material. Single-leaf designs are standard for smaller hatches, while double-leaf designs are required for larger hatches. In addition, aluminum products do not require painting. This is reflected in the warranties available with different products. Product warranties range from ten years to lifetime.

Steel is heavier per square foot than aluminum, and thus steel hatches will be heavier and more difficult to open than aluminum hatches of the same size. However, heavy steel hatch doors may have spring-loaded, hydraulic, or gas openers or other specialized features that help in opening the hatch and in keeping it open.

Many hatches are installed in outdoor areas, often in roadways or pedestrian areas. Therefore, the hatch installed for any given application must be designed to withstand the expected load at that location. Hatches are typically solid to withstand either pedestrian or vehicle loading. Pedestrian loading hatches are typically designed to withstand either 150 or 300 pounds per square feet (psf) of loading. The vehicle loading standard is the American Association of State Highway and Transportation Officials (AASHTO) H-20 wheel loading standard of sixteen thousand pounds over an eight-inch by twenty-inch area. It should be noted that these design parameters are for static loads and not dynamic loads; thus, the loading capabilities may not reflect potential resistance to other types of loads that may be more typical of an intentional threat, such as repeated blows from a sledgehammer or pressure generated by bomb blasts or bullets.

The typical design for a watertight hatch includes a channel frame that directs water away from the hatch. This can be especially important in a hatch on a storage tank because this will prevent liquid contaminants from being dumped on the hatch and leaking through into the interior. Hatches can also be constructed with gasket seals that are air, odor, and gas tight.

Typically, hatches for pedestrian loading applications have hinges located on the exterior of the hatch, while hatches designed for H-20 loads have hinges located in the interior of the hatch. Hinges located on the exterior of the hatch may be able to be removed; thereby allowing intruders to remove the hatch door and access the asset behind the hatch. Therefore, installing H-20 hatches even for applications that do not require H-20 loading levels may increase security, because intruders will not be able to tamper with the hinges and circumvent the hatch this way.

In addition to the location of the hinges, stock hinges can be replaced with heavy duty or security hinges that are more resistant to tampering.

The hatch locking mechanism is perhaps the most important part of hatch security. There are a number of locks that can be implemented for hatches, including

- Slam locks (internal locks that are located within the hatch frame)
- Recessed cylinder locks
- Bolt locks
- Padlocks

Ladder Access Control

Energy sector facilities have a number of assets that are raised above ground level, including electrical substations, pumping stations, gas compression facilities, raised petrochemical tanks, raised piping systems, and roof access points into buildings. In addition, communications equipment, antennae, or other electronic devices may be located on the top of these raised assets. Typically, these assets are reached by ladders that are permanently anchored to the asset. For example, raised petrochemical/water tanks typically are accessed by ladders that are bolted to one of the legs of the tank. Controlling access to these raised assets by controlling access to the ladder can increase security at an energy sector facility.

A typical ladder access control system consists of some type of cover that is locked or secured over the ladder. The cover can be a casing that surrounds most of the ladder or a door or shield that covers only part of the ladder. In either case, several rungs of the ladder (the number of rungs depends on the size of the cover) are made inaccessible by the cover, and these rungs can only be accessed by opening or removing the cover. The cover is locked so that only authorized personnel can open or remove it and use the ladder. Ladder access controls are usually installed at several feet above ground level, and they usually extend several feet up the ladder so that they cannot be circumvented by someone accessing the ladder above the control system.

The important features of ladder access control are the size and strength of the cover and its ability to lock or otherwise be secured from unauthorized access.

The covers are constructed from aluminum or some type of steel. This should provide adequate protection from being pierced or cut through. The metals are corrosion resistant so that they will not corrode or become fragile from extreme weather conditions in outdoor applications. The bolts used to install each of these systems are galvanized steel. In addition, the bolts for each cover are installed on the inside of the unit so that they cannot be removed from the outside.

Locks

A lock is a type of physical security device that can be used to delay or prevent a door, a gate, a window, a manhole, a filing cabinet drawer, or some other physical feature from being opened, moved, or operated. Locks typically operate by connecting two pieces together—such as by connecting a door to a door jamb or a manhole to its casement. Every lock has two modes—engaged (or "locked") and disengaged (or "opened"). When a lock is disengaged, the asset on which the lock is installed can be accessed by anyone, but when the lock is engaged, no access to the locked asset is possible.

Before discussing locks and their applicability it is important to discuss key control. Based on our experience, many energy sector facilities (and others) have no idea how many keys for various site/equipment locks have been issued to employees over the years. Many facilities simply issue keys to employees at hiring with no accountability for the keys upon the employee's departure. Needless to say, this is not good security policy. You can have the best-made locks available installed throughout your facilities, but if you do not have proper key control, you do not have proper security.

Locks are excellent security features because they have been designed to function in many ways and to work on many different types of assets. Locks can also provide different levels of security depending on how they are designed and implemented. The security provided by a lock is dependent on several factors, including its ability to withstand physical damage (i.e., can it be cut off, broken, or otherwise physically disabled) as well as its requirements for supervision or operation (i.e., combinations may need to be changed frequently so that they are not compromised and the locks remain secure). While there is no single definition of the "security" of a lock, locks are often described as minimum, medium, or maximum security. Minimum-security locks are those that can be easily disengaged (or "picked") without the correct key or code, or those that can be disabled easily (such as small padlocks that can be cut with bolt cutters). Higher-security locks are more complex and thus are more difficult to pick or are sturdier and more resistant to physical damage.

Many locks, such as many door locks, only need to be unlocked from one side. For example, most door locks need a key to be unlocked only from the outside. A person opens such devices, called single-cylinder locks, from the inside by pushing a button or by turning a knob or handle. Double-cylinder locks require a key to be locked or unlocked from both sides.

Manhole Intrusion Sensors

Manholes are commonly found in energy sector industrial sites. Manholes are designed to provide access to the underground utilities, meter vaults, petrochemical pumping rooms, and so on, and therefore they are potential entry points to a system. Because many utilities run under other infrastructure (roads, buildings), manholes also provide potential access points to critical infrastructure as well as petrochemical process assets. In addition, because the portion of the system to which manholes provide entry is primarily located underground, access to a system through a manhole increases the chance that an intruder will not be seen. Therefore, protecting manholes can be a critical component of guarding an entire plant site and a surrounding community.

There are multiple methods for protecting manholes, including preventing unauthorized personnel from physically accessing the manhole and detecting attempts at unauthorized access to the manhole.

A manhole intrusion sensor is a physical security device designed to detect unauthorized access to the facility through a manhole. Monitoring a manhole that provides access to a chemical plant or processing system can mitigate two distinct types of threats. First, monitoring a manhole may detect access of unauthorized personnel to chemical systems or assets through the manhole. Second, monitoring manholes may also allow the detection of intruders attempting to place explosive or other destructive (WMD) devices into the petrochemical system.

Manhole Locks

A manhole lock is a physical security device designed to delay unauthorized access to the energy sector facility or system through a manhole.

Radiation Detection Equipment for Monitoring Personnel and Packages

One of the major potential threats facing any critical production facility or system is contamination by radioactive substances. Radioactive substances brought on-site at a facility could be used to contaminate the facility, thereby preventing workers from safely entering the facility to perform necessary water treatment tasks. In addition, radioactive substances brought on-site at an energy sector entity could be discharged into the petrochemical waste system, contaminating the downstream water supply. Therefore, detection of radioactive substances being brought on-site can be an important security enhancement.

Different radionuclides have unique properties, and different equipment is required to detect different types of radiation. However, it is impractical and potentially unnecessary to monitor for specific radionuclides being brought on-site. Instead, for security purposes, it may be more useful to monitor for gross radiation as an indicator of unsafe substances.

In order to protect against these radioactive materials being brought on-site, a facility may set up monitoring sites outfitted with radiation detection instrumentation at entrances to the facility. Depending on the specific types of equipment chosen, this equipment would detect radiation emitted from people, packages, or other objects being brought through an entrance.

One of the primary differences between the different types of detection equipment is the means by which the equipment reads the radiation. Radiation may be detected either by direct measurement or through sampling.

Direct radiation measurement involves measuring radiation through an external probe on the detection instrumentation. Some direct measurement equipment detects radiation emitted into the air around the monitored object. Because this equipment detects radiation in the air, it does not require that the monitoring equipment make physical contact with the monitored object. Direct means for detecting radiation include using a walk-through portal-type monitor that would detect elevated radiation levels on a person or in a package, or by using a hand-held detector, which would be moved or swept over individual objects to locate a radioactive source.

Some types of radiation, such as alpha or low-energy beta radiation, have a short range and are easily shielded by various materials. These types of radiation cannot be measured through direct measurement. Instead, they must be measured through sampling. Sampling involves wiping the surface to be tested with a special filter cloth and then reading the cloth in a special counter. For example, specialized smear counters measure alpha and low-energy beta radiation.

Reservoir Covers

Reservoirs are used to store raw or untreated water for petrochemical processing. They can be located underground (buried), at ground level, or on an elevated surface. Reservoirs can vary significantly in size; small reservoirs can hold as little as a thousand gallons, while larger reservoirs may hold many millions of gallons.

Reservoirs can be either natural or man-made. Natural reservoirs can include lakes or other contained water bodies, while man-made reservoirs usually consist of some sort of engineered structure, such as a tank or other impoundment structure. In addition to the water containment structure itself, reservoir systems may also include associated water treatment and distribution equipment, including intakes, pumps, pump houses, piping systems, chemical treatment and chemical storage areas, and so on.

One of the most serious potential threats to the system is direct contamination of the stored water through dumping contaminants into the reservoir. Energy sector facilities have taken various measures to mitigate this type of threat, including fencing off the reservoir, installing cameras to monitor for intruders, and monitoring for changes in water quality. Another option for enhancing security is covering the reservoir using some type of manufactured cover to prevent intruders from gaining physical access to the stored water. Implementing a reservoir cover may or may not be practical, depending on the size of the reservoir (for example, covers are not typically used on natural reservoirs because they are too large for the cover to be technically feasible and cost effective).

A reservoir cover is a structure installed on or over the surface of the reservoir to minimize water quality degradation. The three basic design types for reservoir covers are

- Floating
- Fixed
- Air-supported

A variety of materials are used when manufacturing a cover, including reinforced concrete, steel, aluminum, polypropylene, chlorosulfonated polyethylene, or ethylene interpolymer alloys. There are several factors that affect a reservoir cover's effectiveness, and thus its ability to protect the stored water. These factors include

- The location, size, and shape of the reservoir
- The ability to lay/support a foundation (for example, footing, soil, and geotechnical support conditions)
- The length of time the reservoir can be removed from service for cover installation or maintenance

- Aesthetic considerations
- Economic factors, such as capital and maintenance costs

For example, it may not be practical to install a fixed cover over a reservoir if the reservoir is too large or if the local soil conditions cannot support a foundation. A floating or air-supported cover may be more appropriate for these types of applications.

In addition to the practical considerations for installation of these types of covers, there are a number of operations and maintenance (O&M) concerns that affect the utility of a cover for specific applications, including how different cover materials will withstand local climatic conditions, what types of cleaning and maintenance will be required for each particular type of cover, and how these factors will affect the cover's life span and its ability to be repaired when it is damaged.

The primary feature affecting the security of a reservoir cover is its ability to maintain its integrity. Any type of cover, no matter what its construction material, will provide good protection from contamination by rainwater or atmospheric deposition, as well as from intruders attempting to access the stored water with the intent of causing intentional contamination. The covers are large and heavy, and it is difficult to circumvent them to get into the reservoir. At the very least, it would take a determined intruder, as opposed to a vandal, to defeat the cover.

Passive Security Barriers

One of the most basic threats facing any facility is from intruders accessing the facility with the intention of causing damage to its assets. These threats may include intruders actually entering the facility, as well as intruders attacking the facility from outside without actually entering it (i.e., detonating a bomb near enough to the facility to cause damage within its boundaries).

Security barriers are one of the most effective ways to counter the threat of intruders accessing a facility or the facility perimeter. Security barriers are large, heavy structures that are used to control access through a perimeter by either vehicles or personnel. They can be used in many different ways, depending on how/where they are located at the facility. For example, security barriers can be used on or along driveways or roads to direct traffic to a checkpoint (i.e., a facility may install Jersey barriers in a road to direct traffic in certain direction). Other types of security barriers (crash beams, gates) can be installed at the checkpoint so that guards can regulate which vehicles can access the facility. Finally, other security barriers (i.e., bollards or security planters) can be used along the facility perimeter to establish a protective buffer area between the facility and approaching vehicles. Establishing such a protective buffer can help in mitigating the effects of the type of bomb blast described above, both by potentially absorbing some of the blast, and also by increasing the "stand-off" distance between the blast and the facility (the force of an explosion is reduced as the shock wave travels farther from the source, and thus the farther the explosion is from the target, the less effective it will be in damaging the target).

Security barriers can be either "active" or "passive." Active barriers, which include gates, retractable bollards, wedge barriers, and crash barriers, are readily movable, and thus they are typically used in areas where they must be moved often to allow vehicles to pass—such as in roadways at entrances and exits to a facility. In contrast to active security barriers, passive security barriers, which include Jersey barriers, bollards, and security planters, are not designed to be moved on a regular basis, and thus they are typically used in areas where access is not required or allowed—such as along building perimeters or in traffic control areas. Passive security barriers are typically large, heavy structures that are usually several feet high, and they are designed so that even heavy-duty vehicles cannot go over or though them. Therefore, they can be placed in a roadway parallel to the flow of traffic so that they direct traffic in a certain direction (such as to a guardhouse, a gate, or some other sort of checkpoint), or perpendicular to traffic such that they prevent a vehicle from using a road or approaching a building or area.

Security for Doorways—Side-Hinged Doors

Doorways are the main access points to a facility or to rooms within a building. They are used on the exterior or in the interior of buildings to provide privacy and security for the areas behind them. Different types of doorway security systems may be installed in different doorways depending on the needs or requirements of the buildings or rooms. For example, exterior doorways tend to have heavier doors to withstand the elements and to provide some security to the entrance of the building. Interior doorways in office areas may have lighter doors that may be primarily designed to provide privacy rather than security. Therefore, these doors may be made of glass or lightweight wood. Doorways in industrial areas may have sturdier doors than do other interior doorways and may be designed to provide protection or security for areas behind the doorway. For example, fireproof doors may be installed in chemical storage areas or in other areas where there is a danger of fire.

Because they are the main entries into a facility or a room, doorways are often prime targets for unauthorized entry into a facility or an asset. Therefore, securing doorways may be a major step in providing security at a facility.

A doorway includes four main components:

- The door, which blocks the entrance. The primary threat to the actual door is breaking or piercing through the door. Therefore, the primary security features of doors are their strength and resistance to various physical threats, such as fire or explosions.

- The door frame, which connects the door to the wall. The primary threat to a door frame is that the door can be pried away from the frame. Therefore, the primary security feature of a door frame is its resistance to prying.
- The hinges, which connect the door to the door frame. The primary threat to door hinges is that they can be removed or broken, which will allow intruders to remove the entire door. Therefore, security hinges are designed to be resistant to breaking. They may also be designed to minimize the threat of removal from the door.
- The lock, which connects the door to the door frame. Use of the lock is controlled through various security features, such as keys, combinations, and so on, such that only authorized personnel can open the lock and go through the door. Locks may also incorporate other security features, such as software or other systems to track overall use of the door or to track individuals using the door, and so on.

Each of these components is integral in providing security for a doorway, and upgrading the security of only one of these components while leaving the other components unprotected may not increase the overall security of the doorway. For example, many facilities upgrade door locks as a basic step in increasing the security of a facility. However, if the facilities do not also focus on increasing security for the door hinges or the door frame, the door may remain vulnerable to being removed from its frame, thereby defeating the increased security of the door lock.

The primary attribute for the security of a door is its strength. Many security doors are four- to twenty-gauge hollow metal doors consisting of steel plates over a hollow cavity reinforced with steel stiffeners to give the door extra stiffness and rigidity. This increases resistance to blunt force used to try to penetrate the door. The space between the stiffeners may be filled with specialized materials to provide fire, blast, or bullet resistance to the door.

The Window and Door Manufacturers Association has developed a series of performance attributes for doors. These include

- Structural resistance
- Forced-entry resistance
- Hinge style screw resistance
- Split resistance
- Hinge resistance
- Security rating
- Fire resistance
- Bullet resistance
- Blast resistance

The first five bullets provide information on a door's resistance to standard physical breaking and prying attacks. These tests are used to evaluate the strength of the door and the resistance of the hinges and the frame in a standardized way. For example, the rack load test simulates a prying attack on a corner of the door. A test panel is restrained at one end, and a third corner is supported. Loads are applied and measured at the fourth corner. The door impact test simulates a battering attack on a door and frame using impacts of two hundred foot-pounds by a steel pendulum. The door must remain fully operable after the test. It should be noted that door glazing is also rated for resistance to shattering and so on. Manufacturers will be able to provide security ratings for these features of a door as well.

Door frames are an integral part of doorway security because they anchor the door to the wall. Door frames are typically constructed from wood or steel, and they are installed such that they extend for several inches over the doorway that has been cut into the wall. For added security, frames can be designed to have varying degrees of overlap with, or wrapping over, the underlying wall. This can make prying the frame from the wall more difficult. A frame formed from a continuous piece of metal (as opposed to a frame constructed from individual metal pieces) will prevent prying between pieces of the frame.

Many security doors can be retrofit into existing frames; however, many security door installations including replacing the door frame as well as the door itself. For example, bullet-resistance per Underwriter's Laboratory (UL) 752 requires resistance of the door and frame assembly, and thus replacing the door only would not meet UL 752 requirements.

Valve Lockout Devices

Valves are utilized as control elements in fuel oil/natural gas and petrochemical process piping networks. They regulate the flow of both liquids and gases by opening, closing, or obstructing a flow passageway. Valves are typically located where flow control is necessary. They can be located in the line or at pipeline and tank entrance and exit points. They can serve multiple purposes in a process pipe network, including

- Redirecting and throttling flow
- Preventing backflow
- Shutting off flow to a pipeline or tank (for isolation purposes)
- Releasing pressure
- Draining extraneous liquid from pipelines or tanks
- Introducing chemicals into the process network
- As access points for sampling process water

Valves may be located either aboveground or belowground. It is critical to provide protection against valve tampering. For example, tampering with a pressure relief valve could result in a pressure buildup and potential explosion in the piping network. On a larger scale, addition of a contaminant or noncompatible chemical substance to the chemical processing system through an unprotected valve could result in the catastrophic release of that contaminant to the general population.

Different security products are available to protect aboveground versus belowground valves. For example, valve lockout devices can purchased to protect valves and valve controls located aboveground. Vaults containing underground valves can be locked to prevent access to these valves.

As described above, a lockout device can be used as a security measure to prevent unauthorized access to aboveground valves located within petrochemical processing systems. Valve lockout devices are locks that are specially designed to fit over valves and valve handles to control their ability to be turned or seated. These devices can be used to lock the valve into the desired position. Once the valve is locked, it cannot be turned unless the locking device is unlocked or removed by an authorized individual.

Various valve lockout options are available for industrial use, including

- Cable lockouts
- Padlocked chains/cables
- Valve-specific lockouts

Many of these lockout devices are not specifically designed for use in the energy sector industry (i.e., chains, padlocks) but are available from a local hardware store or manufacturer specializing in safety equipment. Other lockout devices (for example, valve-specific lockouts or valve box locks) are more specialized and must be purchased from safety or valve-related equipment vendors.

The three most common types of valves for which lockout devices are available are gate, ball, and butterfly valves. Each is described in more detail below.

- Gate valve lockouts—Gate valve lockouts are designed to fit over the operating hand wheel of the gate valve to prevent it from being turned. The lockout is secured in place with a padlock. Two types of gate valve lockouts are available: diameter-specific and adjustable. Diameter-specific lockouts are available for handles ranging from one inch to thirteen inches in diameter. Adjustable gate valve lockouts can be adjusted to fit any handle ranging from one inch to six-plus inches in diameter.
- Ball valve lockouts—There are several different configurations available to lock out ball valves, all of which are designed to prevent rotation of the valve handle. The three major configurations available are a wedge shape for one-inch to three-inch

valves, a lockout that completely covers 3/8-inch to eight-inch ball valve handles, and a universal lockout that can be applied to quarter-turn valves of varying sizes and geometric handle dimensions. All three types of ball valve lockouts can be installed by sliding the lockout device over the ball valve handle and securing it with a padlock.

- Butterfly valve lockouts—The butterfly valve lockout functions in a similar manner to the ball valve lockout. The polypropylene lockout device is placed over the valve handle and secured with a padlock. This type of lockout has been commonly used in the bottling industry.

A major difference between valve-specific lockout devices and the padlocked chain or cable lockouts discussed earlier is that valve-specific lockouts do not need to be secured to an anchoring device in the floor or the piping system. In addition, valve-specific lockouts eliminate potential tripping or access hazards that may be caused by chains or cable lockouts applied to valves located near walkways or frequently maintained equipment.

Valve-specific lockout devices are available in a variety of colors, which can be useful in distinguishing different valves. For example, different-colored lockouts can be used to distinguish the type of liquid passing through the valve (i.e., treated, untreated, potable, petrochemical) or to identify the party responsible for maintaining the lockout. Implementing a system of different-colored locks on operating valves can increase system security by reducing the likelihood of an operator inadvertently opening the wrong valve and causing a problem in the system.

Security for Vents

Vents are installed in some aboveground energy sector storage areas to allow safe venting of off-gases. The specific vent design for any given application will vary depending on the design of the chemical storage vessel. However, every vent consists of an open air connection between the storage container and the outside environment. Although these air exchange vents are an integral part of covered or underground chemical storage containers, they also represent a potential security threat. Improving vent security by making the vents tamper resistant or by adding other security features, such as security screens or security covers, can enhance the security of the entire petrochemical processing system.

Many municipalities already have specifications for vent security at their local chemical industrial assets. These specifications typically include the following requirements:

- Vent openings are to be angled down or shielded to minimize the entrance of surface and/or rainwater into the vent through the opening.

- Vent designs are to include features to exclude insects, birds, animals, and dust.
- Corrosion-resistant materials are to be used to construct the vents.

Visual Surveillance Monitoring

Visual surveillance is used to detect threats through continuous observation of important or vulnerable areas of an asset. The observations can also be recorded for later review or use (for example, in court proceedings). Visual surveillance systems can be used to monitor various parts of production, distribution, or pumping/compressing systems, including the perimeter of a facility, outlying pumping stations, or entry or access points into specific buildings. These systems are also useful in recording individuals who enter or leave a facility, thereby helping to identify unauthorized access. Images can be transmitted live to a monitoring station, where they can be monitored in real time, or they can be recorded and reviewed later. Many energy sector facilities have found that a combination of electronic surveillance and security guards provides an effective means of facility security.

Visual surveillance is provided through a closed-circuit television (CCTV) system, in which the capture, transmission, and reception of an image is localized within a closed "circuit." This is different from other broadcast images, such as over-the-air television, which is broadcast over the air to any receiver within range.

At a minimum, a CCTV system consists of

- One or more cameras
- A monitor for viewing the images
- A system for transmitting the images from the camera to the monitor

Specific attributes and features of camera systems, lenses, and lighting systems are presented in table 9.11.

PETROCHEMICAL AND WASTE MONITORING DEVICES

Earlier it was pointed out that proper security preparation really comes down to a three-legged approach: delay, respond, and detect. The third leg of security, to detect, is discussed in this section.

Sensors for Monitoring Petrochemical and Radiological Contamination

Toxicity tests have traditionally been used to monitor petrochemical wastewater effluent streams for National Pollutant Discharge Elimination System (NPDES) permit compliance. This procedure usually includes pretreatment requirements. That is, the petrochemical facility must treat its waste stream to a certain level of safety to ensure

Table 9.11. Attributes of Camera, Lenses, and Lighting Systems

Camera Systems	
Attribute	*Discussion*
Camera type	Major factors in choosing the correct camera are the resolution of the image required and lighting of the area to be viewed. ▪ **Solid state** (including charge coupled devices, charge priming device, charge injection device, and metal oxide substrate)—these cameras are becoming predominant in the marketplace because of their high resolution and their elimination of problems inherent in tube cameras. ▪ **Thermal**—These cameras are designed for night vision. They require no light and use differences in temperature between objects in the field of view to produce a video image. Resolution is low compared with other cameras, and the technology is currently expensive relative to other technologies. ▪ **Tube**—These cameras can provide high-resolution but burn out and must be replaced after one to two years. In addition, tube performance can degrade over time. Finally, tube cameras are prone to burning images into the tube replacement.
Resolution (the ability to see fine details)	User must determine the amount of resolution required depending on the level of detail required for threat determination. A high-definition focus with a wide field of vision will give an optimal viewing area.
Field of vision width	Cameras are designed to cover a defined field of vision, which is usually defined in degrees. The wider the field of vision, the more area a camera will be able to monitor.
Type of image produced (color, black and white, thermal)	Color images may allow the identification of distinctive markings, while black-and-white images may provide sharper contrast. Thermal imaging allows the identification of heat sources (such as human beings or other living creatures) from low-light environments; however, thermal images are not effective in identifying specific individuals (i.e., for subsequent legal processes).
Pan/tilt/zoom (PTZ)	Panning (moving the camera in a horizontal plane), tilting (moving the camera in a vertical plane), and zooming (moving the lens to focus on objects that are at different distances from the camera) allow the camera to follow a moving object. Different systems allow these functions to be controlled manually or automatically. Factors to be considered in PTZ cameras are the degree of coverage for pan and tilt functions and the power of the zoom lens.
Lenses	
Format	Lens format determines the maximum image size to be transmitted.
Focal length	This is the distance from the lens to the center of the focus. The greater the focal length, the higher the magnification, but the narrower the field of vision.
F number	F number is the ability to gather light. Smaller F numbers may be required for outdoor applications where light cannot be controlled as easily.
Distance and width approximation	The distance and width approximations are used to determine the geometry of the space that can be monitored at the best resolution.

(*continued*)

Table 9.11. *(continued)*

Attribute	*Discussion*
	Lighting Systems
Intensity	Light intensity must be great enough for the camera type to produce sharp images. Light can be generated from natural or artificial sources. Artificial sources can be controlled to produce the amount and distribution of light required for a given camera and lens.
Evenness	Light must be distributed evenly over the field of view so that there are no darker or shadowy areas. If there are lighter vs. darker areas, brighter areas may appear washed out (i.e., details cannot be distinguished) while no specific objects can be viewed from darker areas.
Location	Light sources must be located above the camera so that light does not shine directly into the camera.

Source: U.S. Environmental Protection Agency, *Water and Wastewater Security Product Guide,* http://cfpub.epa.gov/safewater/watersecurity/guide (accessed April 4, 2009).

that downstream wastewater treatment plant processes are not affected by industrial waste products.

Chemical sensors that can be used to identify potential threats to process water and industrial petrochemical wastewater systems include inorganic monitors (e.g., chlorine analyzer), organic monitors (e.g., total organic carbon analyzer), and toxicity meters. Radiological meters can be used to measure concentrations of several different radioactive species.

Monitoring can be conducted using either portable or fixed-location sensors. Fixed-location sensors are usually used as part of a continuous, online monitoring system. Continuous monitoring has the advantage of enabling immediate notification when there is an upset. However, the sampling points are fixed, and only certain points in the system can be monitored. In addition, the number of monitoring locations needed to capture the physical, chemical, and biological complexity of a system can be prohibitive. The use of portable sensors can overcome this problem of monitoring many points in the system. Portable sensors can be used to analyze grab samples at any point in the system but have the disadvantage that they provide measurements only at one point in time.

Radiation Detection Equipment

Radioactive substances (radionuclides) are known health hazards that emit energetic waves and/or particles that can cause both carcinogenic and noncarcinogenic health effects. Radionuclides pose unique threats to source water supplies and petrochemical processing, storage, or distribution systems because radiation emitted from radionuclides in chemical or industrial waste systems can affect individuals through several pathways—by direct contact with, ingestion or inhalation of, or external exposure to the contaminated waste stream. While radiation can occur naturally in some cases due to the decay of some minerals, intentional and unintentional releases of man-made radionuclides into petrochemical water feed or petrochemical processing streams is also a realistic threat.

Threats to energy sector and petrochemical facilities from radioactive contamination could involve two major scenarios. First, the facility or its assets could be contaminated, preventing workers from accessing and operating the facility/assets. Second, the feed water supply could be contaminated. These two scenarios require different threat reduction strategies. The first scenario requires that facilities monitor for radioactive substances being brought on-site; the second requires that feed water assets be monitored for radioactive contamination. While the effects of radioactive contamination are basically the same under both threat types, each of these threats requires different types of radiation monitoring and different types of equipment.

COMMUNICATION INTEGRATION

In this section, those devices necessary for communication and integration of energy sector industrial processing operations, such as electronic controllers, two-way radios, and wireless data communications are discussed. Typically, SCADA systems would also be discussed in this section; however, SCADA was discussed in detail in chapter 7.

In regard to security applications, electronic controllers are used to automatically activate equipment (such as lights, surveillance cameras, audible alarms, or locks) when they are triggered. Triggering could be in response to a variety of scenarios, including tripping of an alarm or a motion sensor; breaking of a window or a glass door; variation in vibration sensor readings; or simply through input from a timer.

Two-way wireless radios allow two or more users that have their radios tuned to the same frequency to communicate instantaneously with each other without the radios being physically lined together with wires or cables.

Wireless data communications devices are used to enable transmission of data between computer systems and/or between a SCADA server and its sensing devices, without individual components being physically linked together via wires or cables. In industrial petrochemical processing systems, these devices are often used to link remote monitoring stations (i.e., SCADA components) or portable computers (i.e., laptops) to computer networks without using physical wiring connections.

Electronic Controllers

An electronic controller is a piece of electronic equipment that receives incoming electric signals an uses preprogrammed logic to generate electronic output signals based on the incoming signals. While electronic controllers can be implemented for any application that involves inputs and outputs (for example, control of a piece of machinery in a factory), in a security application, these controllers essentially act as the system's "brain" and can respond to specific security-related inputs with preprogrammed output response. These systems combine the control of electronic circuitry with a logic function such that circuits are opened and closed (and thus equipment is turned on and off) through some preprogrammed logic. The basic principle behind the operation of an electrical controller is that it receives electronic inputs from sensors or any device generating an electrical signal (for example, electrical signals from motion sensors) and then uses its preprogrammed logic to produce electrical outputs (for example, these outputs could turn on power to a surveillance camera or to an audible alarm). Thus, these systems automatically generate a preprogrammed, logical response to a preprogrammed input scenario.

The three major types of electronic controllers are timers, electromechanical relays, and programmable logic controllers (PLCs), which are often called "digital relays." Each of these types of controller is discussed in more detail below.

Timers use internal signal/inputs (in contrast to externally generated inputs) and generate electronic output signals at certain times. More specifically, timers control electric current flow to any application to which they are connected and can turn the current on or off on a schedule prespecified by the user. Typical timer range (amount of time that can be programmed to elapse before the timer activates linked equipment) is from 0.2 seconds to ten hours, although some of the more advanced timers have ranges of up to sixty hours. Timers are useful in fixed applications that don't require frequent schedule changes. For example, a timer can be used to turn on the lights in a room or building at a certain time every day. Timers are usually connected to their own power supply (usually 120–240 V).

In contrast to timers, which have internal triggers based on a regular schedule, electromechanical relays and PLCs have both external inputs and external outputs. However, PLCs are more flexible and more powerful than are electromechanical relays, and thus this section focuses primarily on PLCs as the predominant technology for security-related electronic control applications.

Electromechanical relays are simple devices that use a magnetic field to control a switch. Voltage applied to the relay's input coil creates a magnetic field, which attracts an internal metal switch. This causes the relay's contacts to touch, closing the switch and completing the electrical circuit. This activates any linked equipment. These types of systems are often used for high-voltage applications, such as in some automotive and other manufacturing processes.

Two-Way Radios

Two way radios, as discussed here, are limited to a direct unit-to-unit radio communication, either via single unit-to-unit transmission and reception or via multiple handheld units to a base station radio contact and distribution system. Radio frequency spectrum limitations apply to all handheld units and are directed by the FCC. This also distinguishes a handheld unit from a base station or base station unit (such as those used by an amateur [ham] radio operator), which operate under different wavelength parameters.

Two-way radios allow a user to contact another user or group of users instantly on the same frequency and to transmit voice or data without the need for wires. They use "half-duplex" communication, or communication that can be only transmitted or received; it cannot be transmitted and received simultaneously. In other words, only one person may talk, while other personnel with radio(s) can only listen. To talk, the user depresses the talk button and speaks into the radio. The audio then transmits the voice wirelessly to the receiving radios. When the speaker has finished speaking and the channel has cleared, users on any of the receiving radios can transmit, either to answer the first transmission or to begin a new conversation. In addition to carrying voice data, many types of wireless radios also allow the transmission of digital data, and these radios may be interfaced with computer networks that can use or track these data. For example, some two-way radios can send information such as global positioning system (GPS) data or the ID of the radio. Some two-way radios can also send data through a SCADA system.

Wireless radios broadcast these voice or data communications over the airwaves from the transmitter to the receiver. While this can be an advantage in that the signal emanates in all directions and does not need a direct physical connection to be received at the receiver, it can also make the communications vulnerable to being blocked, intercepted, or otherwise altered. However, security features are available to ensure that the communications are not tampered with.

Wireless Data Communications

A wireless data communication system consists of two components: a "wireless access point" (WAP), and a "wireless network interface card" (sometimes also referred to as a "client"), which work together to complete the communications link. These wireless systems can link electronic devices, computers, and computer systems together using radio waves, thus eliminating the need for these individual components to be directly connected together through physical wires. While wireless data communications have widespread application in water and wastewater systems, they also have limitations. First, wireless data connections are limited by the distance between components (radio waves scatter over a long distance and cannot be received efficiently unless special directional antennae are used). Second, these devices function only if the individual components are in a direct line of sight with each other, since radio waves are affected by interference from physical obstructions. However, in some cases, repeater units can be used to amplify and retransmit wireless signals to circumvent these problems. The two components of wireless devices are discussed in more detail below.

(1) WAP: The WAP provides the wireless data communication service. It usually consists of a housing (which is constructed from plastic or metal, depending on the environment in which it will be used) containing a circuit board; flash memory that holds software; one of two external ports to connect to existing wired networks; a wireless radio transmitter/receiver; and one or more antenna connections. Typically, the WAP requires a one-time user configuration to allow the device to interact with the local area network (LAN). This configuration is usually done via a Web-driven software application that is accessed via a computer.

(2) Wireless network interface card/client: A wireless card is a piece of hardware that is plugged into a computer and enables that computer to make a wireless network connection. The card consists of a transmitter, functional circuitry, and a receiver for the wireless signal, all of which work together to enable communication between the computer, its wireless transmitter/receiver, and its antenna connection. Wireless cards are installed in a computer through a variety of connections, including USB adapters or laptop CardBus (PCMCIA) or desktop peripheral (PCI) cards. As with the WAP, software is loaded onto the user's computer, allowing configuration of the card so that it may operate over the wireless network

Two of the primary applications for wireless data communications systems are to enable mobile or remote connections to a LAN and to establish wireless communications links between SCADA remote telemetry units (RTUs) and sensors in the field. Wireless card connections are usually used for LAN access from mobile computers. Wireless cards can also be incorporated into RTUs to allow them to communicate with sensing devices that are located remotely.

CYBER PROTECTION DEVICES

Various cyber protection devices are currently available for use in protecting energy sector computer systems. These protection devices include antivirus and pest eradication software, firewalls, and network intrusion hardware/software. These products are discussed in this section.

Antivirus and Pest Eradication Software

Antivirus programs are designed to detect, delay, and respond to programs or pieces of code that are specifically designed to harm computers. These programs are known as "malware." Malware can include computer viruses, worms, and Trojan horse programs (programs that appear to be benign but that have hidden harmful effects).

Pest eradication tools are designed to detect, delay, and respond to "spyware" (strategies that websites use to track user behavior, such as by sending "cookies" to the user's computer) and hacker tools that track keystrokes (keystroke loggers) or passwords (password crackers).

Viruses and pests can enter a computer system through the Internet or through infected floppy discs or CDs. They can also be placed onto a system by insiders. Some of these programs, such as viruses and worms, then move within a computer's drives and files, or between computers if the computers are networked to each other. This malware can deliberately damage files, utilize memory and network capacity, crash application programs, and initiate transmissions of sensitive information from a PC. While the specific mechanisms of these programs differ, they can infect files and even the basic operating program of the computer firmware/hardware.

The most important features of an antivirus program are its abilities to identify potential malware and to alert a user before infection occurs, as well as its ability to respond to a virus already resident on a system. Most of these programs provide a log so that the user can see which viruses have been detected and where they were detected. After detecting a virus, the antivirus software may delete the virus automatically, or it may prompt the user to delete the virus. Some programs will also fix files or programs damaged by the virus.

Various sources of information are available to inform the general public and computer system operators about new viruses being detected. Since antivirus programs use signatures (or snippets of code or data) to detect the presence of a virus, periodic updates are required to identify new threats. Many antivirus software providers offer free upgrades that are able to detect and respond to the latest viruses.

Firewalls

A firewall is an electronic barrier designed to keep computer hackers, intruders, or insiders from accessing specific data files and information on an energy sector's computer network or other electronic/computer systems. Firewalls operated by evaluating and then filtering information coming through a public network (such as the Internet) into the utility's computer or other electronic system. This evaluation can include identifying the source or destination addresses and ports and allowing or denying access based on this identification.

There are two methods used by firewalls to limit access to the utility's computers or other electronic systems from the public network:

- The firewall may deny all traffic unless it meets certain criteria.
- The firewall may allow all traffic through unless it meets certain criteria.

A simple example of the first method is to screen requests to ensure that they come from an acceptable (i.e., previously identified) domain name and Internet protocol address. Firewalls may also use more complex rules that analyze the application data to determine whether the traffic should be allowed through. For example, the firewall may require user authentication (i.e., use of a password) to access the system. How a firewall determines which traffic to let through depends on the network layer at which it operates at and how it is configured. Some of the pros and cons of various methods to control traffic flowing in and out of the network are provided in table 9.12.

A firewall may be a piece of hardware, a software program, or an appliance card that contains both.

Advanced features that can be incorporated into firewalls allow for the tracking of attempts to log onto the local area network system. For example, a report of successful and unsuccessful long-in attempts may be generated for the computer specialist to analyze. For systems with mobile users, firewalls allow remote access into the private network by the use of secure log-on procedures and authentication certificates. Most firewalls have a graphical user interface for managing the firewall.

In addition, new Ethernet firewall cards that fit in the slot of an individual computer bundle additional layers of defense (like encryption and permit/deny) for individual computer transmissions to the network interface function. These new cards have only a slightly higher cost than traditional network interface cards.

Network Intrusion Hardware/Software

Network intrusion detection and prevention systems are software- and hardware-based programs designed to detect unauthorized attacks on a computer network system.

While other applications, such as firewalls and antivirus software, share similar objectives with network intrusion systems, network intrusion systems provide a deeper layer of protection beyond the capabilities of these other systems because they evaluate patterns of computer activity rather than specific files.

It is worth noting that attacks may come from either outside or within the system (i.e., from an insider) and that network intrusion detection systems may be more applicable for detecting patterns of suspicious activity from inside a facility (i.e., accessing sensitive data, etc.) than are other information technology solutions.

Network intrusion detection systems employ a variety of mechanisms to evaluate potential threats. The types of search and detection mechanisms are dependent upon the level of sophistication of the system. Some of the available detection methods include:

- Protocol analysis—Protocol analysis is the process of capturing, decoding, and interpreting electronic traffic. The protocol analysis method of network intrusion detection involves the analysis of data captured during transactions between two or more

Table 9.12. Pros and Cons of Various Firewall Methods for Controlling Network Access

Method	*Description*	*Pros*	*Cons*
Packet filtering	Incoming and outgoing packets (small chunks of data) are analyzed against a set of filters. Packets that make it through the filters are sent to the requesting system and all others are discarded. There are two types of packet filtering: static (the most common) and dynamic.	Static filtering is relatively inexpensive, and little maintenance required. It is well suited for closed environments where access to or from multiple addresses is not allowed.	Leaves permanent open holes in the network; allows direct connection to internal hosts by external sources; offers no user authentication; can be unmanageable in large networks.
Proxy service	Information from the Internet is retrieved by the firewall and then sent to the requesting system and vice versa. In this way, the firewall can limit the information made known to the requesting system, making vulnerabilities less apparent.	Only allows temporary open holes in the network perimeter. Can be used for all types of authentication. Internal protocol services.	Allows direct connections to internal hosts by external clients; offers no user authentication.
Stateful pattern recognition	This method examines and compares the contents of certain key parts of an information packet against a database of acceptable information. Information traveling from inside the firewall to the outside is monitored for specific defining characteristics, and then incoming information is compared to these characteristics. If the comparisons yield a reasonable match, the information is allowed through. If not, the information is discarded.	Provides a limited time window to allow packets of information to be sent; does not allow any direct connections between internal and external hosts; supports user-level authentication.	Slower than packet filtering; does not support all types of connections.

Source: U.S. Environmental Protection Agency, *Water and Wastewater Security Product Guide*, http://cfpub.epa.gov/safewater/watersecurity/guide (accessed April 4, 2009).

systems or devices and the evaluation of these data to identify unusual activity and potential problems. Once a problem is isolated and recorded, problems or potential threats can be linked to pieces of hardware or software. Sophisticated protocol analysis will also provide statistics and trend information on the captured traffic.

- Traffic anomaly detection—Traffic anomaly detection identifies potential threatening activity by comparing incoming traffic to "normal" traffic patterns and identifying deviations. It does this by comparing user characteristics against thresholds and triggers defined by the network administrator. This method is designed to detect attacks that span a number of connections rather than a single session.
- Network honeypot—This method establishes nonexistent services in order to identify potential hackers. A network honeypot impersonates services that don't exist by sending fake information to people scanning the network. It identifies the attacker when he or she attempts to connect to the service. There is no reason for legitimate traffic to access these resources because they don't exist; therefore, any attempt to access them constitutes an attack.
- Anti-intrusion detection system evasion techniques—These methods are designed to thwart attackers who may be trying to evade intrusion detection system scanning. They include methods called IP defragmentation, TCP streams reassembly, and deobfuscation.

While these detection systems are automated, they can only indicate patterns of activity, and a computer administrator or other experienced individual must interpret activities to determine whether or not they are potentially harmful. Monitoring the logs generated by these systems can be time consuming, and there may be a learning curve to determine a baseline of "normal" traffic patterns from which to distinguish potential suspicious activity.

REFERENCES AND RECOMMENDED READING

Garcia, M. L. 2001. *The design and evaluation of physical protection systems.* Burlington, MA: Butterworth-Heinemann.

IBWA. 2004. *Bottled water safety and security.* Alexandria, VA: International Bottled Water Association.

NAERC. 2002. *Security guidelines for the electricity sector.* Washington, DC: North American Electric Reliability Council.

Schneier, B. 2000. *Secrets & lies.* New York: Wiley.

U.S. Environmental Protection Agency. 2005. *Water and wastewater security product guide.* http://cfpub.epa.gov/safewater/watersecurity/guide (accessed April 4, 2009).

10

The Paradigm Shift

Wind turbine, Oklahoma

The 9/11 shift: There is a new world view in the making

The events of 9/11 dramatically changed this nation and focused us on combating terrorism. As a result, in 2003 and subsequent years, the Department of Homeland Security (DHS) in conjunction with members from the general public, state and local agencies, and private groups concerned with the safety of critical infrastructures established the Water Security Working Group (WSWG) to consider and make recommendations on infrastructure security issues. For example, the WSWG identified active and effective security practices for critical infrastructure and provided an approach for adopting these practices. It also recommended mechanisms to provide incentives that facilitate broad and receptive response among critical infrastructure sectors to implement active and effective security practices. Finally, WSWG recommended mechanisms to measure progress and achievements in implementing active and effective security practices and identified barriers to implementation.

The WSWG recommendations on security are structured to maximize benefits to critical industries by emphasizing actions that have the potential both to improve the quality or reliability of service and to enhance security. These recommendations, based on original recommendations from the 2003 National Drinking Water Advisory Council (NDWAC), were designed primarily, as the name suggests, for use by water systems of all types and sizes, including systems that serve fewer than 3,300 people. However, it is the authors' opinion, based on personal experience, that NDWAC's recommendations, when properly adapted to applicable circumstances, can be applied to any and all critical infrastructure sectors, including the energy sector.

The NDWAC identified fourteen features of active and effective security programs that are important to increasing security and relevant across the broad range of utility circumstances and operating conditions. The U.S. Environmental Protection Agency (2003) points out that the fourteen features are, in many cases, consistent with the steps needed to maintain technical, management, and operational performance capacity related to overall water quality; these steps can be applied to other critical infrastructures as well. Many facilities may be able to adopt some of the features with minimal, if any, capital investment.

FOURTEEN FEATURES OF ACTIVE AND EFFECTIVE SECURITY

It is important to point out that the fourteen features of active and effective programs emphasize that "one size does not fit all" and that there will be variability in security approaches and tactics among energy sector facilities, based on industry-specific circumstances and operating conditions. The fourteen features

- Are sufficiently flexible to apply to all chemical industries, regardless of size.
- Incorporate the idea that active and effective security programs should have measurable goals and time lines.
- Allow flexibility for energy sector industrial facilities to develop specific security approaches and tactics that are appropriate to industry-specific circumstances.

Energy sector facilities can differ in many ways, including

- Transportation supply source (rail, air, water, pipeline, lines, or ground)
- Number of supply sources
- Energy processing capacity
- Operation risk
- Location risk
- Security budget
- Spending priorities

- Political and public support
- Legal barriers
- Public versus private ownership

Energy sector industrial facilities should address security in an informed and systematic way, regardless of these differences. Energy sector facilities need to fully understand the specific local circumstances and conditions under which they operate and develop a security program tailored to those conditions. The goal in identifying common features of active and effective security programs is to achieve consistency in security program outcomes among energy sector industrial facilities, while allowing for and encouraging facilities to develop utility-specific security approaches and tactics. The features are based on a comprehensive "security management layering system" approach that incorporates a combination of public involvement and awareness, partnerships, and physical, chemical, operational, and design controls to increase overall program performance. They address industry security in four functional categories: *organization*, *operation*, *infrastructure*, and *external*. These functional categories are discussed in greater detail below.

- **Organizational**—There is always something that can be done to improve security. Even when resources are limited, the simple act of increasing organizational attentiveness to security may reduce vulnerability and increase responsiveness. Preparedness itself can help deter attacks. The first step to achieving preparedness is to make security a part of the organizational culture so that it is in the day-to-day thinking of frontline employees, emergency responders, and management of every energy sector facility in this country. To successfully incorporate security into "business as usual," there must be a strong commitment to security by organization leadership and by the supervising body, such as the board of stockholders. The following features address how a security culture can be incorporated into an organization.
- **Operational**—In addition to having a strong culture and awareness of security within an organization, an active and effective security program makes security part of operational activities, from daily operations, such as monitoring of physical access controls, to scheduled annual reassessments. Energy sector industries will often find that by implementing security into operations they can also reap cost benefits and improve the quality or reliability of the energy service.
- **Infrastructure**—These recommendations advise utilities to address security in all elements of energy sector industry infrastructure—from source to distribution and through processing and product delivery.
- **External**—Strong relationships with response partners and the public strengthen security and public confidence. Two of the recommended features of active end effective security programs address this need.

Fourteen Features

Feature 1. Make an explicit and visible commitment of the senior leadership to security.

Energy sector industrial facilities should create an explicit, easily communicated, enterprise-wide commitment to security, which can be done through

- Incorporating security into a utility-wide mission or vision statement, addressing the full scope of an active and effective security program—that is, protection of worker/public health, worker/public safety, and public confidence, and that is part of core day-to-day operations.
- Developing an enterprise-wide security policy or set of policies.

Energy sector industries should use the process of making a commitment to security as an opportunity to raise awareness of security throughout the organization, making the commitment visible to all employees and customers, and to help every facet of the enterprise recognize the contribution it can make to enhancing security.

Feature 2. Promote security awareness throughout the organization.

The objective of a security culture should be to make security awareness a normal, accepted, and routine part of day-to-day operations. Examples of tangible efforts include

- Conducting employee training
- Incorporating security into job descriptions
- Establishing performance standards and evaluations for security
- Creating and maintaining a security tip line and suggestion box for employees
- Making security a routine part of staff meetings and organization planning
- Creating a security policy

Feature 3. Assess vulnerabilities and periodically review and update vulnerability assessments to reflect changes in potential threats and vulnerabilities.

Because circumstances change, energy sector industrial facilities should maintain their understanding and assessment of vulnerabilities as a "living document" and continually adjust their security enhancement and maintenance priorities. Energy sector industrial facilities should consider their individual circumstances and establish and implement a schedule for review of their vulnerabilities.

Assessments should take place once every three to five years at a minimum. Energy sector industries may be well served by doing assessments annually.

The basic elements of sound vulnerability assessments are

- Characterization of the chemical processing system, including its mission and objectives
- Identification and prioritization of adverse consequences to avoid
- Determination of critical assets that might be subject to malevolent acts that could result in undesired consequences
- Assessment of the likelihood (qualitative probability) of such malevolent acts from adversaries
- Evaluation of existing countermeasures
- Analysis of current risk and development of a prioritized plan for risk reduction

Feature 4. Identify security priorities and, on an annual basis, identify the resources dedicated to security programs and planned security improvements, if any.

Dedicated resources are important to ensure a sustained focus on security. Investment in security should be reasonable, considering utilities' specific circumstances. In some circumstances, investment may be as simple as increasing the amount of time and attention that executives and managers give to security. Where threat potential or potential consequences are greater, greater investment likely is warranted.

This feature establishes the expectation that energy sector facilities should, through their annual capital, operations, maintenance, and staff resources plans, identify and set aside resources consistent with their specific identified security needs. Security priorities should be clearly documented and should be reviewed with utility executives at least once per year as part of the traditional budgeting process.

Feature 5. Identify managers and employees who are responsible for security, and establish security expectations for all staff.

- Explicit identification of security responsibilities is important for development of a security culture with accountability.
- At minimum, energy sector industrial facilities should identify a single, designated individual responsible for overall security, even if other security roles and responsibilities will likely be dispersed throughout the organization.
- The number and depth of security-related roles will depend on a utility's specific circumstances.

Feature 6. Establish physical and procedural controls to restrict access to energy industrial infrastructure to only those conducting authorized, official business and to detect unauthorized physical intrusions.

Examples of physical access controls include fencing critical areas, locking gates and doors, and installing barriers at site access points. Monitoring for physical intrusion can include maintaining well-lighted facility perimeters, installing motion detectors, and utilizing intrusion alarms. The use of neighborhood watches, regular employee rounds, and arrangements with local police and fire departments can support identifying unusual activity in the vicinity of facilities.

Examples of procedural access controls include inventorying keys, changing access codes regularly, and requiring security passes to pass through gates to an access-sensitive area. In addition, utilities should establish the means to readily identify all employees, including contractors and temporary workers, with unescorted access to facilities.

Feature 7. Develop employee protocols for detection of contamination consistent with the recognized limitations in current contaminant detection, monitoring, and surveillance technology.

Until progress can be made in development of practical and affordable online contaminant monitoring and surveillance systems, most energy sector industrial facilities must use other approaches to contaminant monitoring and surveillance.

Many utilities already measure the above parameters (and many others) on a regular basis to control plant operations and confirm chemical mixture quality; more closely monitoring these parameters may create operational benefits for facilities that extend far beyond security, such as reducing operating costs and chemical usage. Energy sector industrial facilities also should thoughtfully monitor customer complaints and improve connections with local public health networks to detect public health anomalies. Customer complaints and public health anomalies are an important way to detect potential contamination problems and other environmental quality concerns.

Feature 8. Define security-sensitive information; establish physical, electronic, and procedural controls to restrict access to security-sensitive information; detect unauthorized access; and ensure information and communications systems will function during emergency response and recovery.

Protecting IT systems largely involves using physical hardening and procedural steps to limit the number of individuals with authorized access and to prevent access by unauthorized individuals. Examples of physical steps to harden SCADA and IT networks include installing and maintaining firewalls and screening the network for viruses. Examples of procedural steps include restricting remote access to data networks and safeguarding critical data through backups and storage in safe places. Utilities should strive for continuous operation of IT and telecommunications systems, even in the event of an attack, by providing uninterruptible power supply and backup systems, such as satellite phones.

In addition to protecting IT systems, security-sensitive information should be identified and restricted to the appropriate personnel. Security-sensitive information could be contained within

- Facility maps and blueprints
- Operations details
- Hazardous material utilization and storage
- Tactical level security program details
- Any other information on utility operations or technical details that could aid in planning or execution of an attack

Identification of security-sensitive information should consider all ways that utilities might use and make public information (e.g., many energy industrial facilities may at times engage in competitive bidding processes for construction of new facilities or infrastructure). Finally, information critical to the continuity of day-to-day operations should be identified and backed up.

Feature 9. Incorporate security considerations into decisions about acquisition, repair, major maintenance, and replacement of physical infrastructure; include consideration of opportunities to reduce risk through physical hardening and adoption of inherently lower-risk design and technology options.

Prevention is a key aspect of enhancing security. Consequently, consideration of security issues should begin as early as possible in facility construction (i.e., it should be a factor in building plans and designs). However, to incorporate security considerations into design choices, energy facilities need information about the types of security design approaches and equipment that are available and the performance of these designs and equipment in multiple dimensions. For example, energy sector facilities would want to evaluate not just the way that a particular design might contribute to security but also how that design would affect the efficiency of day-to-day plant operations and worker safety.

Feature 10. Monitor available threat-level information and escalate security procedures in response to relevant threats.

Monitoring threat information should be a regular part of a security program manager's job, and utility-, facility-, and region-specific threat levels and information should be shared with those responsible for security. As part of security planning, energy sector facilities should develop systems to access threat information, implement procedures that will be followed in the event of increased industry or facility threat levels, and be prepared to put these procedures in place immediately so that adjustments are seamless. Involving local law enforcement and the FBI is critical.

Energy sector facilities should investigate which networks and information sources might be available to them locally and at the state and regional level. If a utility cannot gain access to some information networks, attempts should be made to align with those who can and will provide effective information to the energy sector facility.

Feature 11. Incorporate security considerations into emergency response and recovery plans, test and review plans regularly, and update plans to reflect changes in potential threats, physical infrastructure, processing operations, critical interdependencies, and response protocols in partner organizations.

Energy sector facilities should maintain response and recovery plans as "living documents." In incorporating security considerations into their emergency response and recovery plans, energy sector facilities also should be aware of the National Incident Management System (NIMS) guidelines established by the Department of Homeland Security and of regional and local incident management commands and systems, which tend to flow from the national guidelines.

Energy sector facilities should consider their individual circumstances and establish, develop, and implement a schedule for review of emergency response and recovery plans. Energy sector facility plans should be thoroughly coordinated with emergency response and recovery planning in the larger community. As part of this coordination, a mutual aid program should be established to arrange in advance for exchanging resources (personnel or physical assets) among agencies within a region in the event of an emergency or disaster that disrupts operation. Typically, the exchange of resource is based on a written, formal mutual aid agreement. For example, Florida's Water-Wastewater Agency Response Network (FlaWARN), deployed after Hurricane Katrina, allowed the new "utilities helping utilities" network to respond to urgent requests from Mississippi for help to bring facilities back on line after the hurricane.

The emergency response and recovery plans should be reviewed and updated as needed annually. This feature also establishes the expectation that energy sector facilities should test or exercise their emergency response and recovery plans regularly.

Feature 12. Develop and implement strategies for regular, ongoing security-related communications with employees, response organizations, rate-setting organizations, and customers.

An active and effective security program should address protection of public health, public safety (including infrastructure), and public confidence. Energy sector facilities should create an awareness of security and an understanding of the rationale for their overall security management approach in the communities they reside in and/or serve.

Effective communication strategies consider key messages; who is best equipped/ trusted to deliver the key messages; the need for message consistency, particularly during an emergency; and the best mechanisms for delivering messages and for receiving information and feedback from key partners. The key audiences for communication strategies are: utility employees, response organizations, and customers.

Feature 13. Forge reliable and collaborative partnerships with the communities served, managers of critical interdependent infrastructure, response organizations, and other local utilities.

Effective partnerships build collaborative working relationships and clearly define roles and responsibilities so that people can work together seamlessly if an emergency should occur. It is important for energy sector facilities within a region and neighboring regions to collaborate and establish a mutual aid program with neighboring utilities, response organizations, and sectors, such as the power sector, on which utilities rely or impact. Mutual aid agreements provide for help from other organizations that is prearranged and can be accessed quickly and efficiently in the event of a terrorist attack or natural disaster. Developing reliable and collaborative partnerships involves reaching out to managers and key staff and other organizations to build reciprocal understanding and to share information about the facility's security concerns and planning. Such efforts will maximize the efficiency and effectiveness of a mutual aid program during an emergency response effort, as the organizations will be familiar with each others' circumstances and thus will be better able to serve each other.

It is also important for energy sector facilities to develop partnerships with the communities and customers they serve. Partnerships help to build credibility within communities and establish public confidence in utility operations. People who live near energy sector facility structures can be the eyes and ears of the facility and can be encouraged to notice and report changes in operating procedures or other suspicious behaviors.

Energy sector facilities and public health organizations should establish formal agreements on coordination to ensure regular exchange of information between facilities and public health organizations and outline roles and responsibilities during response to and recovery from an emergency. Coordination is important at all levels of the public health community—national public health, county health agencies, and health-care providers, such as hospitals.

Feature 14. Develop energy facility–specific measures of security activities and achievements, and self-assess against these measures to understand and document program progress.

Although security approaches and tactics will be different depending on energy utility–specific circumstances and operating conditions, we recommend that all energy sector facilities monitor and measure a number of common types of activities and achievements, including existence of program policies and procedures and training, testing, and implementing schedules and plans.

The Fourteen-Feature Matrix

In table 10.1, a matrix of recommended measures to assess effectiveness of an energy sector facility's security program is presented. Each feature is grouped according to its functional category: organization, operation, infrastructure, and external.

Table 10.1. Fourteen Features of Active and Effective Security Matrix

Features	*Checklist: Potential Measures of Progress*
Organizational Features	
Feature 1—Explicit commitment to security	Does a written, enterprise-wide security policy exist, and is the policy reviewed regularly and updated as needed?
Feature 2—Promote security awareness	Are incidents reported in a timely way, and are lessons learned from incident responses reviewed and, as appropriate, incorporated into future utility security efforts?
Feature 5—Defined security roles and employee expectations	Are managers and employees who are responsible for security identified?
Operational Features	
Feature 3—Vulnerability assessment up to date	Are reassessments of vulnerabilities made after incidents, and are lessons learned and other relevant information incorporated into security practices?
Feature 4—Security resources and implementation priorities	Are security priorities clearly identified, and to what extent do security priorities have resources assigned to them?
Feature 7—Contamination detection	Is there a protocol/procedure in place to identify and respond respond to suspected contamination events?
Feature 10—Threat-level based protocols	Is there a protocol/procedure of responses that will be made if threat levels change?
Feature 11—Emergency response plan tested and up to date	Do exercises address the full range of threats—physical, cyber, and contamination—and is there a protocol/procedure to incorporate lessons learned from exercises and actual response into updates to emergency response and recovery plans?
Feature 14—Industry-specific measures and self-assessment	Does the utility perform self-assessment at least annually?
Infrastructure Features	
Feature 6—Intrusion detection and access control	To what extent are methods to control access to sensitive assets in place?
Feature 8—Information protection and continuity	Is there a procedure to identify and control security-sensitive sensitive information, is information correctly categorized, and how do control measures perform under testing?

Features	*Checklist: Potential Measures of Progress*
Feature 9—Design and construction standards	Are security considerations incorporated into internal utility design and construction standards for new facilities/ infrastructure and major maintenance projects?
External Features	
Feature 12—Communications	Is there a mechanism for utility employees, partners, and the community to notify the utility of suspicious occurrences and other security concerns?
Feature 13—Partnerships	Have reliable and collaborative partnerships with customers, managers of independent interrelated infrastructure, and response organizations been established?

Source: U. S. Environmental Protection Agency, *Active and Effective Water Security Programs*, http://cfpub.epa.gov/safewater/watersecurity/features.cfm (accessed June 2006).

Ultimately, the goal of implementing the fourteen security features (and all other security provisions) is to create a significant improvement in the energy sector industry on a national scale by reducing vulnerabilities and therefore risk to public health from terrorist attacks and natural disasters. To create a sustainable effect, the energy sector as a whole must not only adopt and actively practice the features, but also incorporate the features into "business as usual."

REFERENCE

U.S. Environmental Protection Agency. 2003. *Active and effective water security programs.* http://cfpub.epa.gov/safewater/watersecurity/features.cfm (accessed June 2006).

Index

About the Authors

Frank R. Spellman is Assistant Professor of Environmental Health at Old Dominion University. He is a professional member of the American Society of Safety Engineers, the Water Environment Federation, and the Institute of Hazardous Materials Managers. He is also a Board Certified Safety Professional and Board Certified Hazardous Materials Manager with more than 35 years of experience in environmental science and engineering. He is the author of more than fifty books, including *Water Infrastructure Protection and Homeland Security* (GI, 2007) and *Food Supply Protection and Homeland Security* (GI, 2008).

Revonna M. Bieber is currently working for the Naval Medical Center Portsmouth in the field of industrial hygiene. Her work focuses on environmental health hazards and radiologic and health-care safety. She is the co-author, with Frank R. Spellman, of *Chemical Infrastructure Protection and Homeland Security* (GI, 2009) and numerous other books.